The Dawnseekers

Robert West Howard

The Dawnseekers

THE FIRST HISTORY OF AMERICAN PALEONTOLOGY

Foreword by Gilbert F. Stucker
The American Museum of Natural History

HBJ

New York and London
HARCOURT BRACE JOVANOVICH

Printed in the United States of America

Library of Congress Cataloging in Publication Data

Howard, Robert West, 1908–
The dawnseekers.

Bibliography: p.
Includes index.
1. Paleontology—United States—History. I. Title
QE705.U6H68   560'.973   74-30407
ISBN 0-15-123973-8

First edition

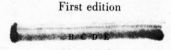

To

Dr. Paul McGrew—for suggesting and
implementing the idea.

Drs. George P. Lovell and Wheelock Southgate—
for keeping me alive long enough to research and
write the book.

Dr. Eleanore Larson—for a magic mixture of
intellectual and emotional gifts.

Contents

Illustrations

Foreword

A chapter of American history lies buried in the voluminous annals of paleontology, the science of ancient life. It is a dramatic story of the search for truth hidden in the stony layers of the earth—of discoveries made and the men who made them, scientists of heroic mold who not only gave us our knowledge of the dawn eras of time, but played crucial roles in shaping our national character.

Eight years ago, Robert West Howard set himself the task of backtracking this trail into the past. It turned into a formidable undertaking—a paleontological feat of its own, for it was a rediscovering of what had long remained outside the public consciousness. At one point, Bob's health failed, but not his determination. The result is *The Dawnseekers*.

The book is the first comprehensive approach to the broad, historical aspects of American paleontology. By happy circumstance, its publication comes at a time of sharply rising interest in the backgrounds of the geological sciences among the professionals themselves. A committee of eminent American earth scientists has been organized to further this interest. Members of the Geological Society of America give it increasing attention at their annual meetings, and the U.S. Geological Survey plans a popular history of its organization to commemorate its centennial in 1979.

Bob Howard is a master in matters historical. He writes with authority, yet with freshness and the feel of life. He does more than excavate the dry bones of fact and figure; he resurrects them, fleshes them out, gives them living dimension.

As he traces the course of the science of fossils from the early mastodon bogs of New York's Wallkill valley to the elaborate, latter-day surveys and dinosaur digs of the Rocky Mountain West, a picture emerges of men driven by a hunger of the spirit. Whatever other considerations may have moved them, it was the lure of the unknown that drew the dawnseekers like a magnet into those virginal, barren regions where stratified rocks held the remains of creatures so long dead they had turned to stone.

It was the petrified mysteries preserved in the ancient sea beds of New York state that Amos Eaton pursued when surveying the projected route of the Erie Canal. These fossils provided the subject matter for his lectures to the farmers and villagers along the way, as he developed the teaching techniques that would later influence the founding of Rensselaer Polytechnic Institute in Troy, "the nation's first institution for technical and science training." It was the promise of bone that beckoned Hayden and Meek westward, to find, in the Big Badlands of what is now South Dakota, the world's richest source of fossil mammals. It was the hope of discovering extinct forms new to science that sent the rival teams of Cope and Marsh scurrying across the continental divide to the lush deposits of the Green River Basin.

As the book describes, the probings reached northward into the broken country along the upper Missouri and Judith rivers, southward to the Smoky Hill River in Kansas and the arroyos of New Mexico. The first freewheeling expeditions gave way to the highly organized, government-funded territorial surveys, which were later combined to form the present U.S. Geological Survey. Where the requisites of coal, iron, soil, water, and the like were found, farms, industries, and communities took root. Areas of surpassing beauty or scientific interest, many of them first described in the dawnseekers' field reports, became our national parks and monuments.

The fossil treasures, crated to the centers of learning in the East, triggered intellectual ferment, nurtured new thinking.

They gave comparative anatomist Joseph Leidy the material for his epochal monograph, *The Extinct Mammalian Fauna of Dakota and Nebraska* (1869), which elevated the study of American fossil vertebrates to the status of an officially recognized science. More significantly, as Bob Howard points out, they confirmed Darwin's theory of evolution, released to a stunned world in 1859 in *The Origin of Species*. They were conclusive evidence that the earth and its inhabitants were not the products of a single act of creation during "one week in 4004 B.C.," as strict interpreters of the Bible maintained. Creation was continuous through all the ages of time—a growing out of the past, through present into future. Genesis? Genesis was here and now!

Perhaps no man realized this truth and its implications for humanity more than Henry Fairfield Osborn, who declared that, on the basis of the fossil record, "Evolution was firmly grounded as a Law of living Nature." It was through his efforts that the endeavors of the dawnseekers reached their fullest expression. As president of the American Museum of Natural History from 1908 to 1933 and founder of its Department of Vertebrate Paleontology, he brought those endeavors to new levels of productivity and gave them world-wide scope. Balancing exploration, research, public exhibition, and publication, he welded the science into a unified, organic whole.

A bust stands in a hall of the museum, bearing his likeness. The legend beneath it reads, "For him the dry bones came to life and giant forms of ages past rejoined the pageant of the living."

For us, it is the dawnseekers themselves that have been brought back to life, in this remarkable book by Bob Howard.

Gilbert F. Stucker

The American Museum of Natural History
New York, New York
October 3, 1974

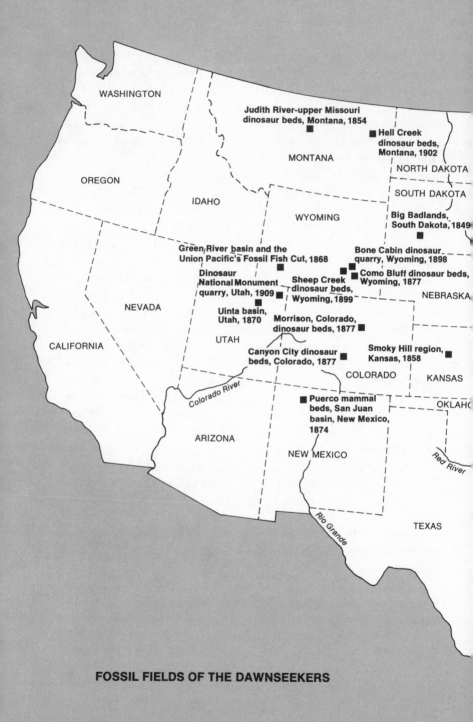

FOSSIL FIELDS OF THE DAWNSEEKERS

The Dawnseekers

1 | The Reverend Annan's Ominous Molars

We know, on the authority of Moses, that longer ago than six thousand years, the world did not exist.—Martin Luther

AN understandable patriotism and an unfortunate pedantry have held the body of American history in military and political stays. More often than not such seemingly exotic Americana as Dr. Christian Friedrich Michaelis's epochal dig for *Mammut* teeth during the fall of 1782 in New York's Wallkill valley is missed. The big things in history often occur in a silence that strikes later generations with palpable force. Such is the case with what the good doctor's determined and energetic curiosity uncovered. The sequence of pioneering during the next century, in fact, testifies that the Michaelis dig was a paleontological Pandora's box. Tumbling out of it came the mute evidence which aroused the curiosity and determined the research that transformed the *natural philosopher* into the modern *scientist* and eventually uncovered enough about Earth's vast age and substance to produce the industrial plunder and technological plenty of the entrepreneurial century from 1870 to 1970, as well as the contemporary potential of ecologic catastrophe.

The first American museums, the first university courses in geology and soil science, the first schools of engineering, the angry schisms of Protestant sects, the awesome birth of paleontology and the industries that resulted from it were all large steps toward the achievement of an American civilization. They were as meaningful—and as heroic—as the wagon trains, gold rushes, Indian wars, and homesteading. The Philosophers of Philadelphia, the Peales, Benjamin Silliman, Amos Eaton,

Joseph Leidy, F. V. Hayden, and Othniel Marsh stand as tall on
the horizon of American folk heroes as Daniel Boone, Davy
Crockett, Lewis and Clark, Zebulon Pike, and Jedediah Smith.
In searching for data about the dawn epochs of our planet, they
blazed the trails to the minerals industries, the strip mines, the
oil refineries, the atomic bombs, and the polluted lakes. Yet
they were as innocent of deadly intent in this search for dawn
facts as the Chinese who first touched a glowing ember to an
experimental mixture of sulphur, charcoal, and saltpeter.

The dictionaries provide, as we shall see, abundant paleontological testimony of their own about the social and intellectual
convictions that dominated the Atlantic coastal plain when a
minority of the residents of the Thirteen Colonies rebelled
against George III and England's Parliament. For more than a
thousand years, Christians had accepted the clergy's decision
that Earth was a young and unchanging planet that would probably be destroyed "very soon" by a divinely ordered conflagra-
tion, with a coincidental sorting out of the "eternally damned"
and the heaven-bound "meek." Since 1650, both Catholics and
Protestants had accepted Bishop Ussher of Ireland's research
and his conclusion that Earth's Creation had been completed on
Sunday, October 10, 4004 B.C. Except for a few terrain changes
made during the forty days of Noah's Flood, the Ussher research assured, everything on Earth was precisely as the Almighty's will created it during 144 day and night hours in 4004
B.C.

This was the general conception of past, present, and future
when the minutemen stood their ground at Concord's bridge,
when the young man from Charlottesville, Virginia, wrote out
the final draft of the Declaration of Independence, and when a
massive French fleet off Virginia's York peninsula persuaded
General Charles Cornwallis that he had to surrender his command to "those damned ragatag Continentals."

During April, 1782, the thousands of Loyalists who had
taken refuge in New York City and Brooklyn began planning
mass migrations to Nova Scotia and New Brunswick. General
Washington and several thousand Continental troops were in

barracks at New Windsor, near the Hudson valley village of Newburgh. Save for the probability of a few terrorist raids by Joseph Brant's Iroquois warriors and such Loyalist fanatics as the green-coated Tories led by John and Walter Butler, an uneasy truce settled across the United States from the Kennebec to the Savannah.

The more reason, thus, for glum suspicions when a packet boat delivered a strange request from British headquarters in New York City. The letter had been brought upstream by a British sloop delivering routine correspondence about prisoners of war and the formalities of reuniting members of Loyalist and Continental families via "flag" escorts through the Westchester wilderness. But this letter requested permission for a Dr. Christian Friedrich Michaelis to visit the farm of a Reverend Annan, fifteen miles across the highlands in the Wallkill valley, in order to examine "giant teeth" that Annan had allegedly dug up.

The records do not, of course, detail the discussions between Colonel Alexander Hamilton and other aides before the request was brought to General Washington's attention. But the uncertainties of the truce, the political feuding in the Continental Congress, and the black hate of the Loyalists at Niagara all suggest that there was a determined inquiry about Dr. Michaelis and his real reasons for this prowl to the Wallkill.

Brant and the Butlers were still very capable of organizing a raid southeast across the lake plains and the Susquehanna wilderness to the Wallkill. Was Dr. Michaelis going out to meet them in order to co-ordinate their movements with Loyalist forces skulking up from New Jersey? The objective might be the assassination of General Washington; most Continental veterans were convinced that "if anything happens to the General, those chicken-livered bastards in Congress will surrender to the first redcoat they can find." (Only a few weeks before, Washington had sternly reprimanded the delegation of officers determined to revolt against the Congress and enthrone him as king of America.)

The gruff Prussian drillmaster Baron von Steuben was in his

quarters at New Windsor. Perhaps the aides asked him about the Michaelis family of Göttingen. Perhaps the Baron frowned, tugged at his mustache, then recalled that there was a Professor Michaelis up there somewhere in the ore-rich mountains of Saxony who was a specialist in Oriental history and had a passion for collecting unusual rocks.

And, before the request was advanced to the desk of General Washington's secretary, Intelligence had to be consulted for any scraps of rumor or fact it might possess about either Dr. Michaelis or the Reverend Annan.

The zany experiments that Dr. Michaelis was conducting at the Hessians' hospital on the outskirts of Brooklyn had caused so much tavern gossip that Intelligence had a dossier on him. Rattlesnakes and pit adders—or copperheads, as New Yorkers called them—were so numerous on the Manhattan rocks and in the Brooklyn marshlands that every farm boy knew how to apply a tourniquet, suck the poison out of a bite wound, then paint the area with "Injun herbals." Snakes and the efficiency of their venom fascinated Dr. Michaelis. So he had built a pen in the yard behind his barrack quarters, offered a twopenny reward for any poisonous snake delivered alive to him, and then undertook a year-long study of the social and reproductive habits of rattlesnakes and copperheads. One of the rattlers bit him; he experienced the efficiency of both the venom and the Indian herbals. After analyzing the snake year from spring spawn to winter hibernation, it was said, Dr. Michaelis killed the snakes, dissected them, and sent an exhaustive report off to his father in Göttingen.

The ripe cucumber odor of the copperheads still lingered in the pens when Dr. Michaelis began rounding up stray dogs. Through this experiment Michaelis hoped to find a way to alleviate the tortures then essential to leg or arm amputations. Every naval and land battle harvested as many amputees as corpses. Whisky and rum were the only known painkillers, and gangrene was a common sequel to a flesh wound. So, in a valiant effort to learn more about the network of nerves and possible methods for anesthetizing or regenerating them, Dr.

Michaelis began a year of experiments with stray dogs. (The papers he wrote about these experiments were also mailed to Göttingen and also evoked parental praise. Several of them were published in German, Dutch, and British medical journals during 1784–85 and, authorities say, accord Dr. Michaelis the status of pioneer researcher in this area of internal medicine.)

It seemed apparent, then, that Dr. Michaelis was far too intrigued by the mysteries of natural philosophy to be involved in any Loyalist skulduggery.

As for the Reverend Robert Annan, there was abundant evidence that he had discovered a wagonload of weird bones and teeth while digging an irrigation ditch through his Wallkill swampland. Indeed, Annan had ridden into camp with a saddlebag full of the things one afternoon and requested permission to show the whole lot to Washington and formally present him with one of the teeth. Each tooth was as large as a big man's fist. The beast they came from must have stood twenty-five to thirty hands high; twice as big as any known breed of horse.

General Washington had been too busy settling rows between his division commanders and the French admirals, and with pleas to Congress for some of the wages grossly overdue his regiments. He sent his regrets to the Reverend Annan, promised that one day he would ride over to inspect the "giant animal's grave," and inquired about Annan's familiarity with the graveyard of giant animals at Big Lick, Kentucky. One of the Big Lick teeth was a prized memento in his study at Mount Vernon. During the years before the war the mysterious events responsible for those giant bones at Big Lick had formed the core of many hearthside arguments in inns at Williamsburg and Philadelphia as well as at Mount Vernon.

Most New Englanders, of course, clung stubbornly to Cotton Mather's pronouncement that these objects proved that a race of "sinful giants" had been wiped out by Noah's Flood. Ben Franklin conceded that the teeth and bones were evidence of a mighty race that had become extinct, possibly because of some "cataclysmic change in Nature." But Thomas Jefferson, a de-

voted student of natural philosophy, insisted that Almighty God
would never permit a species to become extinct; creatures of
this size, he believed, still wandered in the Far West wilderness.

So with a grunt of assent from Intelligence, Dr. Michaelis's
request to visit the Reverend Annan's farm reached Washington's desk and promptly won his consent.

Thunderheads were swathing Storm King, Mount Beacon,
and the other highland crags in silver-trimmed black cloaks
when a packet sloop landed the young physician at the New
Windsor wharf. Officers at the Fort Montgomery frontier had
already riffled through his baggage. Two of the cleverest youngsters in Intelligence were at the wharf to escort him to quarters
and, since the impending storm ruled that the trip to the Wallkill be postponed until the next day, to determine the validity of
such a harebrained project as a bone hunt.

They managed to keep their expressions calm and not to
choke on their hot buttered rum that evening when Dr. Michaelis admitted that "old bones" had been his principal reason
for volunteering as a physician with the Hessian mercenaries
George III had hired to squelch the American rebellion.

Christian Friedrich Michaelis was the son of Johann David
Michaelis, whose curiosity about mankind's history came peril-
ously close to heresy. Johann became obsessed with evidence of
ancient life after examining some of the bones and gigantic leaf
imprints that miners were discovering in the ore and coal pits of
Saxony. He had studied the writings of the great metallurgist
Georg Agricola, who had used the same digs two hundred years
before as the laboratory for his masterwork, *De Re Metallica*.

Agricola contended, and Johann Michaelis heartily concurred, that the Golden Fleece sought by the Argonauts of the
Greek myth was actually an uncured sheep pelt used to pan gold
flakes and small nuggets on the eastern shore of the Black Sea.
When mud and gravel were sluiced over the pelt, the gold
adhered to the lanolin in the fleece. After the pelt was heavy
with gold, it was sun-dried and then burned, and the gold was
recovered. The Argonaut legend, Agricola concluded, was a

poetic account of an actual piracy, based on the earliest known method of panning for gold.

If, Professor Michaelis reasoned, the Argonaut saga was the embroidered cloak for a metallurgic truth and a bit of economic plundering, then why couldn't there be other historical truths hidden in other sagas? Could it be, as a few valiant doubters whispered, that the "seven days and nights" of Creation in the Bible's Genesis were also symbolism? Was there an Earth before 4004 B.C.? The evidence coming in from both the Siberian and American wildernesses demanded research. Huge ivory tusks had become commonplace barter at the trading posts on Russia's Don and Volga rivers. The Vikings who had established these posts centuries before assumed that the tusks were being passed along the ancient Silk Road out of India. But now Orientalists at Saint Petersburg and Paris were expressing doubts. Huge tusks continued to show up at trade fairs each summer; most of them were twice as large as any elephant tusk ever seen in Europe, and they were not white but a soft golden color. It took centuries to impart such a patina to ivory. The suspicion grew that the Mongol nomads of Siberia were plundering the graveyards of giant animals somewhere near the Arctic Ocean. But why were there elephant graveyards in the Arctic?

Then, during the 1760's, molars and pieces of tusks that were as large and seemed as ancient as the Russian ivory reached Paris and London. Some, forwarded from New England, were identified merely as "from the New York wilderness." Others had been found in a swampy meadow at Big Lick, on the Ohio River.

Christian Michaelis inherited his father's curiosity and zest for research. He studied medicine at the University of Strasbourg, and in 1776 went to London for postgraduate studies and internship. It was there that he learned that the rebellion in the American colonies was more than a street riot. For the next two years he spent his free afternoons studying the bones of the "American Monsters" exhibited at the British Museum, and he

sent sketches and descriptions of them to his father. Both father
and son became so intensely curious about the bone graveyards
and their terrain that they decided Christian must somehow get
to America to investigate the sites firsthand. The opportunity
came in 1779 when the British government hired more troops
from the princes of Hesse. Johann arranged for his son's
appointment as a resident physician with a Hessian regiment.

Christian had been stationed in Brooklyn for three years,
waiting for the day when the rebels would be chased far enough
inland to permit a visit to the Big Lick graveyard. When word
of the Reverend Annan's discovery reached him, he realized it
might be his only opportunity to visit a giants' graveyard—and
now, here he was.

The young physician sighed and reached for his mug of rum.
It had grown tepid during the long explanation; its beads of
butter gleamed gold in the firelight. A dozen officers had
crowded into the cabin to hunker along the walls. His official
hosts, from Intelligence, were staring fixedly into the fire and
seemed pleased by whatever vision they saw there.

"Dammit, Doctor," a voice snapped from the shadows, "you
should be given a pass to Philadelphia."

"Philadelphia?" Christian turned. "I do not understand."

"Just what I was thinking, sir," said one of the Intelligence
youngsters. "There is an organization in Philadelphia, Doctor,
called the American Philosophical Society. Many of its members share your curiosity about bones, rocks, and other mysteries of nature."

"And," continued the voice in the shadow, "they have a
damned fine collection of those teeth from Big Lick."

"Of course," drawled another man, "most of the clergy and
not a few physicians will assure you that these are fragments
from skeletons of the animals who died on Noah's Ark and were
buried at sea."

"More sick critters on that ship," a third man mused. "Can't
understand why ol' Noah didn't know more 'bout tendin' livestock. Gotta admit, though, that was a right smart boat. Just
forty days, but she tossed out carcasses from Gibraltar to

Pennsylvania and then all the way back to Mount Ararat. John Paul Jones coulda used a craft that pert, 'stead of them rotten hulks the Frenchies give 'im."

There was a snigger, then a silence. Then the first man spoke again: "Man you should see in Philadelphia, Doctor, is an artist named Peale. Won't let you take away any of those bones from the Philosophical Society, I'm sure. But Peale could do fine drawings of them to take home to your pa."

"We'll discuss the matter with headquarters while Doctor Michaelis is on the Wallkill," said one of the Intelligence youngsters. He rose and stretched. "A most interestin' evening, Doctor. Pleasant journey and . . . *gut nacht.* Is that correct?"

"*Guten nacht. Auf wiedersehen.*"

The cart road between New Windsor and the Wallkill valley had been built for the red-white-and-blue Conestoga wagons hauling iron ore from Bull Mine to the Greenwood and Cornwall furnaces, and so had been vital in the production of cannons and bar iron for the Continental army. Its twin ruts ran in gentle switchbacks up the Hudson's bluffs through butternut and hickory groves to hushed emerald forests of hemlock and white pine. East of them, the Hudson River was a fat pewter ribbon arcing between the bluffs of Livingston Manor to the north and the blue-black tusks of Storm King to the south, intricately hemstitched between by baby-blue threads of smoke rising from fireplaces, cookhouses, and mills.

After a half hour's jog across a meadow studded with granite and iron-stained sandstone boulders, the traveler found the terrain tilting down toward the Wallkill's boggy shores. To Christian Michaelis, the panorama recalled a headland on the English Channel. Less than a hundred feet wide, the Wallkill looped through bog and black prairie that ranged from two hundred yards to a mile in width. To the west, the valley sloped sharply upward through terraced headlands composed of layers of varicolored granite, sandstone, and basalt that were stacked as neatly as giant flapjacks. Each stratum was contoured as subtly and cleanly as the Dover cliffs along the English Channel.

When the Reverend Annan learned that the group approach-

ing his property included that young Hessian doctor he had
been briefed about a few days before, he excitedly bellowed the
news up to the house. He led the way to the bone pile in the
carriage house, apologizing that the heap kept getting smaller
and smaller because he couldn't resist visitors' pleas for "a
little piece to take home for the folks to see." But sections of leg
bone "as big as a fireplace backlog," crumbling ribs and
vertebrae, as well as four coal-black teeth, still remained.

After the guests had been introduced to the Annan family,
settled their belongings in the rooms over the kitchen, and en-
joyed a two-hour lunch of brown trout, corn bread, sweets and
sours, and deep-dish apple pie, Annan took Christian out
along the lane to the wheat field where the bones had been
discovered. He held forth in his loud bass voice about neighborhood theories, similar discoveries across the Hudson in
Livingston Manor, and the grandiloquent conclusions of the
Reverend Cotton Mather. Christian grunted and nodded now
and then while he silently mourned the irreparable damage that
the Reverend Annan and others like him did, despite their
good intentions. Probably the grave beside the Wallkill had contained a fairly complete skeleton of this elephantlike giant. A
physician, or even a farmer who was reasonably familiar with
animal bone structure, would have removed the earth carefully
from the carcass, drawn sketches, and taken measurements.
Only then would he have removed the pieces—one at a time.
But Annan and his neighbors seemed to have jumped in like
children in a strawberry patch. Bones had been whacked,
crushed, and tossed helter-skelter. Chemical stains and possibly
flesh remnants were trampled into the muck. The best Christian
could hope for was additional information about the terrain,
perhaps a few pieces off that bone pile . . . or extraordinary
luck on the part of the ten infantrymen Colonel Hamilton had
promised to send over from New Windsor to serve as his dig
crew.

The pit where the bones had lain was trampled and crushed
into a quagmire. The earth was oily black loam with veins of
sand and clay crisscrossing it. At one time there had been

quicksand here. The creature had very likely wandered into a quicksand pocket and died there, thrashing and bellowing. If that were the case, there must be other ancient quicksand traps in this valley. With luck . . . perhaps . . .

The squad of diggers arrived on a creaking Conestoga rig the next morning. The black-and-silver cloak hung over Storm King again. The thunder echoed and the hail banged like buckshot while the crew gossiped and pitched halfpennies on the barn floor. Rain drummed on the cedar shingles all night.

By morning the Wallkill was in flood; the meadows were too wet to dig for anything but worms. Christian rowed across to the west shore, determined to cross the prairie and climb those terraced bluffs. He thought the view from the crest might give him some clues as to the most promising places for digging. But he was just as anxious to examine more closely the black, red, and gray layers of rock.

An hour later he was clambering up the black layer. It was slick and fine-grained. Suddenly he spotted something glittering on the ledge face a few feet away. He slithered over, brushed away a layer of moss, and studied it. Firmly embedded in the rock face was a jet-black object about two inches long and an inch wide that resembled a gigantic bedbug, with dozens of spidery legs arcing out from the oval body. He tried to pry it loose, but couldn't.

Thirty feet up, the rock abruptly changed to a sandstone so rich in iron that it was almost apple red. He searched for another specimen of the jet-black object, but found none. But some of the sandstone was studded with small shells that were fan-shaped and fluted like scallops. In the crumbly gray shale near the crest there were other shell-shaped objects that looked like small oysters and clams.

Agricola had written about similar shell and bug shapes found hundreds of feet beneath the surface in the Saxony mines. There was a collection of them on one of the shelves in his father's study.

What was it the clergymen thundered about from their pulpits? The mysterious ways of God? And the whimsical ways of

God? Why had He played such mischievous pranks that fateful October week in 4004 B.C.?

The rains lasted a full week. The Wallkill muck turned so boggy that the sergeant of the dig crew grumbled, "It's like spoonin' pea soup, sir."

Christian Michaelis sighed and gave the order to start packing. He wangled a giant tooth and some bone splinters away from the Reverend Annan. Perhaps Colonel Hamilton would be able to secure permission for him to visit Philadelphia. The physicians and natural philosophers in the American Philosophical Society would certainly have more information about the giant animals and their habits than this back-country parson could provide. And if Peale was half as skilled an artist as Colonel Hamilton said he was, he should be able to produce a set of drawings that would enable Christian's father to make fairly accurate comparisons of American, Siberian, and Saxon mystery bones.

The rain was so persistent that Christian lashed his horse's reins to the Conestoga's tail gate and crawled in among the dig crew for the ride back over the highland spur to New Windsor. He huddled down with arms wrapped around his knees and eyes fixed on the Wallkill's terraced west shore. Wisps of fog drifting along the crest became white tracings on the wall of an ancient temple. He mused again about that gleaming black bug fixed in the lowest stone layer, the scallop shells in the red, the coarser oyster shapes and clam pieces in the shale at the top.

The glory of God in His mysterious whimsies! How many generations had pondered about these mysteries? Would men one day dare to challenge the edicts of the clergy and search for a more plausible explanation of God's will? If they did, this valley would play a role in the search.

He didn't know precisely why, he just knew it would.

2 | The Heretics

The whole vital process of the Earth takes place so gradually and in periods of time which are so immense compared with the length of our life, that these changes are not observed. Before their course can be recorded from beginning to end, whole nations perish and are destroyed.—Aristotle, *Meteorologica.*

Discussions about the vast age of our planet, its sequence of life forms, and the comparative insignificance of that Johnny-come-lately, the human being, echoed morning after morning of 341 and 340 B.C. in the gardens of Pella, Macedonia. Determined that his daredevil heir become as skilled in logic and earth lore as he was in horsemanship and warfare, King Philip had lured Plato's most gifted pupil, Aristotle of Stagira, north from Athens to tutor Alexander and his companions.

Neither priests nor royal counselors divulged to Philip that Alexander, Craterus, and the other young noblemen were being deceived by Artistotle's pragmatic details about an Earth not only ancient beyond comprehension but in continuing change. They too were eagerly auditing the discussions and inviting the teacher to their villas to continue the talk through the afternoon repose and evening feasts. Both the cliffs of nearby Mount Olympus and the beaches of the Aegean gleamed with shells and other fossil evidence of Earth's vast age and persistent change. A century and a half before, Xenophanes of Colophon had taught that the impressions of seashells and ferns found in rocks were, like the agatized teeth and giant bones washed up on the beaches, evidence that violent change had occurred time and time again during Earth's past. These, Xenophanes was certain, reached so far back in the planet's history that man could neither measure, nor comprehend, the ages since Creation.

Aristotle's lectures about what we now call *paleontology* so impressed Alexander that discussions of state religions and regional lore became a routine part of Alexander's epochal conquests from the Hellespont to India. Summaries of Alexander's audiences with priests and philosophers were sent back to Aristotle in Athens along with specimens of minerals, rare plants, and strange animals the Macedonians found in Egypt, Persia, and India. Thus, in Chaldea, Alexander cross-examined priests about their conviction that Earth had "emerged from chaos" more than two million years before. At Babylon, local astrologers were equally certain that the Earth had been "born" 500,000 years ago. Across the red deserts, in the Persian heartland, priests believed the planet was quite young and had been formed a mere twelve thousand years before.

But beyond the snow thrones of the Hindu Kush, in India, the Brahmans preached a philosophy of humility surprisingly similar to Aristotle's. Earth and time, they claimed, were "infinite," and this tiny planet, which man's slowly swelling brain had recently enabled him to dominate, was too ancient and complex for that brain's comprehension. Because each stage of the Earth's development from fire through ice to habitable climes covered a vast period of time, the Brahmans divided the distant past into "days" and "nights." Each "day" was an epoch of land emergence; each "night" was a titanic span of earthquakes, volcanic terrors, and glaciers.

Back on the Mediterranean shore, the small kingdom of the Israelites had evolved a state religion based on the premise of an Almighty God who had been responsible for the creation of Earth as well as the heavens. The sacred records of this religion began as chants recited, or sung *a capella*, by tribal elders. Since the Israelites were active traders and in constant contact with people from other lands, it is not unlikely that they borrowed folklore from neighboring cultures and transferred such tales as the building of the Ark and the revelation of the Ten Commandments to their own folk heroes. These stories, passed from generation to generation through recitation and chant, evolved over the centuries into rich poetic narratives. The account of Jehovah's creation of Earth and the human being, as finally recorded on clay tablets and the splendid new papyrus, was exquisitely direct poetry, with imagery that was as stimulating to children and simple shepherds as it was to sages and priests.

Somewhere back in forgotten centuries, the composers of this Book of Genesis learned the succinct terms "day" and "night" that the Brahman composers of the Veda had invented to denote the vast eras of the Earth's dawn history. And "seven" had long been a number deemed to have unique magic properties. Logically, thus, Genesis recounted that Jehovah created Earth in six "days and nights," then rested on the supermagical seventh "day."

To give the religion as firm a base as possible, the scholars of Israel deliberated and finally agreed on October 7, 3761 B.C. (here adapted to the modern calendar), as "the final day of Creation."

When Christianity, after three centuries of being regarded as a "heretic cult," was proclaimed by Emperor Constantine as the new state religion, its leaders agreed that the Hebraic account of Creation should be retained as the first book of the faith's most sacred text, the Holy Bible. Clement of Alexandria, Theophilus of Antioch, and other leading theologians accepted the Hebrew scholars' conclusion that Creation must have occurred about thirty-eight centuries before Christ's birth.

In their effort to carry their religion to as many parts of the world as possible, the Christians revised their doctrinaire account to incorporate some of the nature worshipers' most important symbols: the pine tree, the mistletoe, the rites and festivals surrounding spring planting, the autumn harvest, and the winter solstice. But a different strategy became necessary to explain the strange objects that kept challenging the date of Creation.

Each spring—or so it seemed to the harried priests—plowmen turned up scores of arrowheads, daggers, and axes that had been neatly chipped from obsidian or flint. Peat cutters dug up the bones, tusks, and teeth of huge animals "never seen in these parts." Pigs rooting for truffles snouted up chunks of pottery intricately painted with scenes of naked men and women engaging in blatantly polygamous sex acts. Quarry slaves and miners found imprints of giant leaves and ferns, skeletons of fish, and petrified shells in rocks far below the earth's surface and hundreds of miles from the ocean. Winter gales tumbled nodules of amber up on the Baltic and Aegean beaches; some of the lovely chunks contained the fossil carcasses of flies, wasps, and small land animals.

How could these things exist, the more reflective Christians asked, if the Earth was only four or five thousand years old?

Christianity had already splintered into rival sects. Now there was danger of further schisms, for the objects that were being unearthed cast doubt on the agreed-on date of Creation or perhaps even implied that the Almighty had been careless in permitting such random dumpings during those six days and nights of Creation. It seemed necessary to bring such supernatural beings into the folklore as witches, ghosts, elves, and trolls.

When, about 1000 A.D., the Eastern emperor in Constantinople sent the Western emperor in Rome an obsidian battle-ax that had been dug up from a cave floor near the Black Sea, the scroll accompanying it explained that the object was a "Heaven Ax of the type used during the war between Satan and God." A century later, in Brittany, the bishop of Rennes ruled that the numerous flint and bronze tools being plowed up by the peasants were "divinely appointed means of securing success in battle, safety on the sea, security against thunder and immunity from unpleasant dreams." Thereafter, Stone Age and Bronze Age tools and weapons became "sacred objects"; scholars and priests called them *thunderstones*.

By the fourteenth century, the church had formulated its own explanation for the artifacts and fossils. Seashells on mountaintops and giant bones were all, the clergy insisted, the remains of animals that had been wiped out by the Great Flood. As for the "thunderstones," they were still being deposited willy nilly by "the Forces of Darkness [in order] to deceive mankind and tempt him from the Path of Righteousness." In 1649, the

French philosopher Tollius explained that "thunderstones are generated in the sky by a fulgorous exhalation conglobed in a cloud by the circumposed humour."

Once these explanations were put forth, some vigorous churchmen insisted on enforcing their acceptance. At Rome, in about 1580, Pope Gregory XIII ordered publication of *The Roman Martyrology*, which declared that the creation of man occurred 5,199 years before Christ. Even religious leaders who broke away from the Catholic church maintained their convictions about recent Creation. Martin Luther insisted that the Earth was not more than six thousand years old, and he held to the belief that all fossils and inland seashells were products of the Great Flood. In England, Sir Matthew Hale based his *Primitive Origination of Man* on the premise of recent Creation. And in 1650, James Ussher, archbishop of Armagh, published his *Annals of the Ancient and New Testaments*. It was the product of decades of research into the development of the Bible through Hebrew, Greek, Latin, and English translations. Earth and man, the archbishop concluded, were created in 4004 B.C., and the task had been completed on October 10. Ussher's data was so impressive, and his status as Hebraic scholar so lofty, that his date of 4004 B.C. was ordered imprinted in the margin of all authorized editions of the Protestants' King James Bible. Soon the Freemasons were dating their official correspondence by adding 4,004 years to the current Christian year, and identifying it with the initials A.L. (for *Anno Lucis*, in the Year of Light). Thus Columbus discovered America in A.L. 5496.

After 1200, any European who challenged the concept of recent Creation was automatically a heretic, subject to religious and political persecution. In 1600, Giordano Bruno was burned at the stake in Rome because he lectured that the seas had drowned the continents "many times."

In the fifteenth century, the Florentine genius Leonardo da Vinci daringly wrote that "the waters of the Deluge could not have carried the live shell animals on its crest and thus to the tops of mountains. The oysters were fastened to the bottom of the sea, while cockles could not have traveled from the Adriatic to the mountains of Lombardy in forty days, since their rate of travel is only three or four braccia a day." Normally, such a statement would have been cause for imprisonment. But Leonardo's royal masters protected him as "a very imaginative fellow of artistic temperament." He escaped with a tart lecture from the bishop of Florence, who pointed out: "The Almighty created Earth as He saw fit. Since He obviously wished to place fishbones and shells in mines and inland cliffs, He must have had reason for it. His Will is not to be questioned."

The French scholar La Peyrère had no princely protector. In 1650, he cited the age of Egyptian and Greek civilizations and Leonardo's arguments about oysters and cockles as justification for his contention that both Earth and man existed before 4000 B.C. The grand vicar of the archdiocese of Mechlin sent soldiers to arrest him and to seize and burn all copies of his treatise. La Peyrère was starved and tortured until he recanted and confessed that his heresy was the result of "Protestant influences."

But Sir Walter Raleigh suffered similarly in Protestant England. Raleigh too was devoted to "the examination of nature," and the wonders he observed on his voyages converted him to a bluff pragmatism. "Reason," he wrote, "must be elicited from fact." In 1585, he sent the artist John White and the scientist Thomas Harriot to Virginia's Roanoke Island to study and report on the terrain and the native folkways in order to prepare for the colony Raleigh planned to establish there. White was commissioned to produce water-color sketches of the Indians, depicting their villages and their social and religious routines. Harriot was to make botanic, zoologic, and topographic studies of the Virginia coastal plain and islands; then, if the Indians could be persuaded to guide him, to explore the mysterious "blue mountains" that filled the western horizon.

John White's paintings and Harriot's book, *A Brief and True Report of the New Found Land of Virginia* (1588), were important contributions to European knowledge about the wilderness that would become the American colonies, and they helped to increase Raleigh's interest in assembling plausible facts for his *History of the World*.

Thomas Harriot became a frequent guest at Raleigh's estate near Sherborne on the Dorset Downs. The lusty playwright-poet Christopher Marlowe and the poet George Chapman occasionally rode down from London too. During these long, and sometimes brawling, holidays, the four argued about the logic of contemporary historians, mathematicians, and natural philosophers. Harriot, who owned a Latin translation of Aristotle, found more "pure reason" in Aristotle's convictions about the Earth's age than he did in Judeo-Christian doctrine. Marlowe too believed that mysteries such as the ruins at Stonehenge and the ancient civilizations being discovered in the Orient and the South Pacific were exciting evidence of a world far older than fifty-six centuries.

Stimulated by the discussions at Raleigh's estate, Harriot formulated mathematical tables to estimate how long it would take natural forces to build up the current density of salt in the waters of the North and Irish seas.

Gossip about the strange goings-on at Sherborne spread outside Dorset. In 1592, the Jesuit Robert Parsons preached vehemently against "Rawleigh's school of athiesm." In 1593, Marlowe was stabbed to death during a tavern argument about "faith in the Bible." In 1594, Queen Elizabeth grudgingly approved the appointment of a commission to investigate "the sett of athiests in Dorset."

As a result of the commission's report, Raleigh was brought to London and given a severe lecture, despite the queen's conviction that Parsons's sermons were Papist intrigue. Raleigh was made governor of Jersey in 1600, and Harriot's visits to Dorset ended. Harriot became a recluse and devoted the rest of his life to mathematics and astronomy—he introduced some of the now standard symbols of algebra, discovered spots on the sun, and recorded the patterns of Jupiter's satellites.

But the taint of "athiesm" continued to hang over Raleigh; it was cited during his trial for treason in 1603. He began his *History of the World* while imprisoned in the Tower of London.

One passage in the book so enraged the king of Spain that he urged James I to execute Raleigh immediately. There was proof positive of heresy, the Spanish ambassador reported, in Raleigh's allegation that "in Abraham's time, all the then known parts of the world were developed. . . . Egypt had many magnificent cities, and these were not built with sticks but of hewn stone. This magnificence needed a parent of more antiquity than these other men have supposed."

The 1603 "treason" sentence was reactivated, and Raleigh was executed on October 29, 1618.

The white settlers who arrived to colonize the New World in the sixteenth century found that, just as in Europe, mysterious bones of huge proportions had been dug up by the natives of North and South America. The Indians had their own beliefs about the origins of the bones.

"The Tlascalans," Bernal Díaz del Castillo recalled in 1568, "said that their ancestors had told them that in former times the country [Mexico] was inhabited by men and women of great stature, and wicked manners, whom their ancestors had at length extirpated; and that in order that we might judge of the bulk of these people, they brought us a bone which had belonged to one of them, so large that when placed upright it was as high as a middling sized man. It was the bone between the knee and the hip.

"They also brought pieces of other bones of great size, but much consumed by time. But the one I have mentioned was entire. We were astonished at these remains, and thought they certainly demonstrated the former existence of giants. This bone was sent to Castille for His Majesty's inspection, by the first persons who went on our affairs from hence."

Folk tales about gigantic humans and animals and drastic climate changes were woven into the religious rituals of tribes from Peru to the Arctic. The Chitimacha, who inhabited the Gulf Coast bayous in what is now Alabama, believed that "a long time ago a being with a long nose came out of the ocean and began to kill people. It would root up trees with its nose to get at people who sought refuge in the branches." The folklore of the Penobscots on the Maine coast told of "great animals with long teeth . . . who were so big that they appeared to be . . . moving hills without vegetation." They related their ancestors' struggles to survive during the years of "the great ice walls" and the tribal hunts for "monstrous moose with a fifth leg."

In the forests of what is now New York state, the Cayuga nation of the Iroquois Confederacy revered a nine-foot-long ivory tusk that they believed had magical powers. It had washed up onto a riverbank during a springtime flood. The Cayugas named the river for the magic tusk: the Chemung, or Big Horn.

But the white clergy held fast to their belief in recent Creation and spoke of dreadful curses of "eternal hellfire and damnation" for those who embraced other theories. The fossil teeth and bones that gleamed on beaches, the giant leg bone shown to Díaz, the Cayugas' magical tusk were dismissed as "objects planted by Satan" or as "evil creatures destroyed during Noah's Flood."

It was in this spirit that the Reverend Cotton Mather, in 1706, concluded that the six-inch teeth and fist-size chunks of leg bone forwarded to him by Massachusetts Governor Joseph Dudley were "remains of godless giants drowned in Noah's Flood." The first American member of England's Royal Society for Improving Natural Knowledge, a Harvard graduate, and associate pastor of Boston's North Church, the Reverend Mather was New England's prime authority on natural history. (His studies of "Satanic possession" and its causes also established him as an authority on witchcraft; in the 1690's he had encouraged the witchcraft trials and hangings at Salem and along the North Shore of Massachusetts Bay.)

The Reverend Mather wanted "the patronage of some generous Maecenas" to finance research and publication of his *Magnalia Christi Americana,* a work intended to prove that the natural history of New England demonstrated God's will as it was "revealed" in Genesis and other books of the Old Testament. He sent the teeth and the bone fragments to the Royal Society, explaining that they had been discovered on a Hudson valley farm in New York colony and seemed to come from the same species of "giant" as a seventeen-foot-long thighbone that had been sent to him from across the Berkshires a few years before.

His judgment that the teeth were those of "a wicked giant" was in accord with the teachings of most natural philosophers of the time. The French scholar Henrion had recently published a study demonstrating that Adam must have been 123 feet 9 inches tall and Eve a petite 118 feet 9 inches. During 1726, mountaineers brought the fossil of a prehistoric salamander down from a gorge for examination by Germany's distinguished natural philosopher Scheuchzer; the professor ruled it to be "another human witness to Noah's Deluge."

The teeth and bones Mather sent were added to the collection of thunderstones, gigantic shins, and Siberian ivory tusks in the Royal Society's cabinet. No encouragement was given to his appeal for financial support.

Since Mather considered himself an authority on supernatural forces, it was fitting that his request was turned down by the spiritual and intellectual heirs of Francis Bacon. Between 1610 and 1627, Bacon had written *Sylva Sylvarum, Historia Ventorum, Novum Organum, The New Atlantis,* and other treatises on natural philosophy and its socioeconomic implications. They were the most cogent works on the subject yet produced by an Englishman. While skillfully avoiding the blunt allegations that had brought accusations of atheism against Sir Walter Raleigh just a few years earlier, Bacon challenged mankind to intensive "examinations of nature" and "experimentation before theory," because "Man, being the servant and interpretare of Nature, can do and understand so much, and so much only, as he had observed, in fact or in thought, of the course of Nature; beyond this he neither knows anything nor can do anything."

Prophetically, in *The New Atlantis,* Bacon envisioned an island in the South Pacific that would be governed by economists, physicians, and philosophers. Its capital city would be enhanced by huge libraries, gardens, zoos, laboratories, and astronomical observatories. He gave precise, detailed descriptions of undersea ships, machine-driven wagons, airplanes, anesthetics, radios, microscopes, hybrid crops and flowers, perpetual-motion clocks, and malleable, brilliant metals that would be developed through "examination of Nature" and pragmatic "experimentation before theory" in this New Atlantis.

Bacon's pleas for "reason which is elicited from facts . . . a *closer* and purer league between the experimental and the rational" took sturdy root at the universities at Oxford, Cambridge, and Edinburgh, especially among physicians and medical students. His work was cited as the inspiration for the founding, in 1660, of the Royal Society for Improving Natural Knowledge. Twenty-eight-year-old Christopher Wren, then a professor of astronomy at Gresham College, London, composed the draft for the Society's charter. Liberally borrowing from Bacon's philosophy, Wren wrote that the Society was to "prosecute effectively the Advancement of Natural Experimental Philosophy especially those parts of it which concern the Encrease of Commerce by the Addition of useful Inventions tending to the Ease, Profit or Health of our Subjects . . . to confer about the hidden Causes of Things . . . and to prove themselves real Benefactors to Mankind."

Although recent Creation and the clergy's catchall explanations for the existence of fossils and artifacts were not openly challenged during the Society's first decades, the "natural experimental philosophy" set forth in its papers and discussions closely reflected Bacon's ideas. Thomas Burnet presented mathematical calculations that demonstrated that the Earth was endlessly changing and that wind, rain, and tide would eventually level all the mountains and wash them into the seas. Robert Hooke reiterated Leonardo's theories about mountaintop seashells and elicited gasps from several members when he suggested that the "objects in the rocks" might be the remains of ancient plants and animals.

The Royal Society's adherence to Bacon's belief in "experimentation before theory" was strong enough for its members to reject the Reverend Mather's plea for an underwriting. They were not so much denying his zeal as upholding the principles on which the Society had been founded.

The members of the Royal Society were not the only people pursuing Bacon's ideals in the early eighteenth century. In Pennsylvania, the Quakers were also attempting to place "experimentation before theory" in exploring their intricate relationships with, and dependence on, the "storehouse of Nature."

The valleys and coastal plains of William Penn's colony contained wide meadowlands; the hardwood forests were rich with limestone soil. Attracted by these natural resources and by the air of intellectual freedom that prevailed in the colony, thousands of refugees from the Palatinate regions of Germany emigrated to Pennsylvania in the early 1700's. These new settlers were the descendants of the farmers who, centuries earlier, had perfected the horse collar, the bank barn, crop field manuring, and other imaginative techniques that aided in the spread of diversified agriculture in Europe. When the settlers arrived in Pennsylvania, discussions and experiments preceded decisions to enforce cropland fencing, specified seed mixtures for pastures, a system of fallowing for grain fields, and fertilization of fields with winter-cured animal manures.

In keeping with this tradition that encouraged curiosity and open discussion, a young, Boston-born refugee from Puritan piety named Benjamin Franklin took over *The Pennsylvania Gazette* in 1730 and began publishing *Poor Richard's Almanack* in 1732. A shrewd publicist, Franklin knew that a richer reading program would not only help the circulation of his publications but would also hasten the development of schools and co-operative efforts to improve the agriculture and social life of the colony. In 1731, he organized America's first book club, the Library Company of Philadelphia, and opened a reading and circulation room in Pewter Platter Alley, in Philadelphia. When the membership rose to one hundred, the club moved to quarters in the brick structure that would become Independence Hall. Franklin and James Logan chose the books and, consciously or not, held to Bacon's ideals by limiting the lists to scholarly works in geography, chemistry, and natural philosophy, and such classics as Tacitus's *Annals*, Plutarch's *Lives*, translations of Aristotle, and the Oxford Dictionary.

But, as Franklin realized, reading stimulates the mind to formulate opinions. Opinions stimulate arguments and, now and again, produce a luxuriant harvest of worth-while ideas. He organized a club, called the Junto, for book discussions. By 1743, the arguments at Junto meetings had become so scholarly that Franklin and Logan reorganized the club into a society for natural philosophy; they named it the American Philosophical Society.

That same year, the English naturalist Mark Catesby published his *Natural History of Carolina, Florida and the Bahamas* in London. It significantly reported that "all parts of Virginia . . . abound in Fossil Shells of various kinds, which in stratums lie imbedded a great Depth in the Earth, in the Banks of Rivers and other Places, among which are frequently found the Vertibras and other Bones of Sea Animals. At a place in Carolina called Stono, was dug out of the Earth three or four teeth of a large Animal which, by the concurring Opinion of all the Negroes, native Africans, that saw them were the Grinders of an Elephant, and in my opinion they could be no other; I having seen some of the like that are brought from Africa."

By the time Catesby's two-volume work reached Franklin's library, reports about another discovery of giant teeth, this time in the Ohio wilderness, must have reached Philadelphia. During the summer of 1739, a canoe expedition commanded by young Baron de Longueuil paddled and portaged from the new French forts at Niagara Falls down the Allegheny and Ohio rivers. Although their official mission was to negotiate a peace treaty with the Chickasaw Indians in Tennessee, the embossed lead plates the baron ceremoniously buried along the route announced that France intended to claim the two-thousand-mile arc from the Saint Lawrence valley to Louisiana's Gulf Coast as New France. This would restrict British North America to a narrow strip of Atlantic coastal plain.

The revelation of France's plans caused so much turmoil in London and the colonial capitals that spies were ordered to

learn all they could about Longueuil's expedition. Thus politicians and military experts as well as natural philosophers
pondered reports about an ivory tusk, three giant teeth, and
chunks of bone recently placed in the Cabinet du Roi at
Versailles.

All of these remains had been shipped from New Orleans to
Paris by Longueuil in 1740. Indian guides, it seemed, had
shown him a "sacred field" somewhere in the Ohio valley where
hundreds of huge bones, teeth, and tusks surrounded a salt
spring. French naturalists pronounced them "remarkably like
an elephant's" and apparent cousins of the ivory tusks being
traded out of Siberia.

The Hudson valley . . . Stono, Carolina . . . and now
"somewhere on the Ohio." An American elephant? Was it
extinct? Or did it still roam somewhere in the wilderness west
of the Alleghenies?

In 1762, when the French and Indian Wars were drawing to
an end and Conestoga wagon drivers had conquered the Allegheny barrier to the Ohio's floatway and Kentucky, a Pennsyl-
vania-born botanist named John Bartram began corresponding
with Fort Pitt traders about "giant bones . . . at or near the
Shawnee towns in the Ohio's mid-valley."

The traders identified the bone yard as "a salt lick of 30 to
40 acres extant, four days journey below the lower Shawanese
towne" and alleged that they had seen "bones scattered here
and there, some upon the surface, some partly burned and all
much decay'd by time."

Up the Great Lakes and west through the Ohio forests, Indian
leaders campaigned for the social freedoms they had repeatedly been promised once "the French tyranny is defeated."
In 1765, the Irish trader George Croghan, recently appointed
Sir William Johnson's deputy on Indian Affairs, took a flotilla
of Durham boats down the Ohio to attempt peace talks. The
expedition reached the Big Bone Lick salt marsh, between
Covington and Louisville, Kentucky, on May 30.

"We came into a large road which the buffaloes have beaten,"
Croghan wrote in his log. "It is spacious enough for two wagons

to go abreast and leading straight into the Lick. It appears that there are vast quantities of these bones lying five or six feet underground, which we discovered in the bank at the edge of the Lick. We found here two tusks about six feet long. We carried one, with some other bones to our boats and set off."

But a week later, near the mouth of the Wabash, Indians ambushed the brigade, captured Croghan, killed five of his companions, destroyed the boats, and confiscated the bones.

Croghan was ransomed and back in Fort Pitt before winter. He organized another trade-and-council flotilla and headed back west on the spring floods of 1766. The party included Ensign Thomas Hutchins, the military engineer who would become the United States' first official geographer, and George Morgan, a Philadelphia trader. The flotilla reached Big Bone Lick on July 17. Croghan had received requests from Philadelphia to secure at least two lots of teeth, tusks, and bone fragments. Both were to be shipped to London: one to Lord Shelburne, His Majesty's minister for the colonies, and one to Benjamin Franklin, resident agent for the Pennsylvania colony. Ensign Hutchins was to survey and prepare a detailed map of the mysterious bone yard. Morgan also wanted some specimens for his brother John, codirector of the College of Medicine at the College of Philadelphia.

This time Croghan evaded ambush attempts, and the bones reached London and Philadelphia during 1767. Dr. John Morgan devoted several evenings to examinations of the items he had received and then placed them in the cabinet of natural curiosities he had collected during a tour of Italy a few years before. He decided neither to clean nor label any of the Big Lick specimens until he could find time for a thorough analysis.

The Big Lick artifacts shipped to London stimulated both research and fiery debate. Lord Shelburne turned his specimens over to the Royal Society, where the contents were studied by Dr. William Hunter and Peter Collinson, the Quaker merchant responsible for John Bartram's recent appointment as botanist-collector of His Majesty's Botanical Gardens. Collinson reported his analysis of the bones and teeth at the November 26 and December 10 meetings of the Royal Society. The tusks, he believed, were remarkably similar to the tusks of contemporary elephants in Africa and Asia. But the teeth differed, and therefore must "belong to another species of elephant not yet known which is probably supported by browsing on trees and shrubs and other vegetable foods."

This explanation did not satisfy Dr. Hunter. London's leading obstetrician, Dr. Hunter, a Scot, was as dourly individualistic as those Edinburgh rebels agitating for a "free church." After analyzing the Big Lick tusks and comparing them with tusks from Siberia and Africa, he had asked London jewelers to examine the ivory.

His research, he told the Society's February 25, 1768, meeting, had convinced him that the giant creatures who had died at Big Lick belonged to a variety of elephant that was carnivorous and "thank heavens . . . is probably extinct."

The intimation that the Earth's Creator was so careless—or whimsical—as to permit a species of carnivorous elephants to become extinct seemed to be flagrant heresy. But Bacon's plea for "experimentation before theory" was receiving increasing support. Peter Collinson did not challenge Hunter's findings; the paper was published the following autumn in the Royal Society Proceedings.

On January 31, 1768, three and a half weeks before Dr. Hunter delivered his paper, Benjamin Franklin sent a request to the Abbé Chappe d'Auteroche in Paris. Its phrasing suggests that Franklin's speculations were also pointing toward the "extinct species" theory, and, moreover, that he too suspected an association between the Big Lick specimens and the ivory tusks coming out of Siberia.

"I sent you sometime since," Franklin wrote, ". . . a Tooth that I mention'd to you when I had the Pleasure of meeting with you at the Marquis de Courtanvaux's. It was found near the River Ohio in America . . . at what is called the Great Licking Place. . . . Some of our Naturalists here contend that these are not the grinders of Elephants but of some carnivorous Animal unknown, because such Knobs of Prominances on the

Face of the Tooth are not to be found on those of Elephants and only, as they say, on those of carnivorous Animals. But it appears to me that Animals capable of carrying such large heavy Tusks, must themselves be large Creatures, too bulky to have the Activity necessary for pursuing and taking Prey, and therefore I am inclined to think those Knobs are only a small Variety [i.e., variation]. . . . I should be glad to have your opinion and to know from you whether any of the kind have been found in Siberia."

One of the foremost astronomers of France, the Abbé Chappe had journeyed to Siberia in 1761 to observe the transit of Venus. Intrigued by the displays of ivory tusks in the market places, he joined a trader's caravan to the interior, saw deposits of giant bones along river beds, and undertook the extraordinary journey home via Alaska and California. The Tartars of the Siberian steppes called the huge elephantlike skeletons *mammots*, which was their word for "giants."

The Abbé Chappe was less impulsive than Dr. Hunter. After a cautious examination of the Big Lick tooth, he wrote Franklin that although there was some similarity between the Siberian and Big Lick molars, both undoubtedly came from creatures who had been drowned during Noah's Flood; the carcasses in America must have been washed there when the Flood ebbed.

Extinct elephants? River valleys created by erosion? Where was fact? Had there been enough experimentation to merit a theory?

The clouds of war moved the pieces of *mammot* back into the cabinets. That's where Dr. John Morgan's boxful had been throughout the Collinson-Hunter-Franklin-Chappe inquiries. They were still there, crusted with Big Lick dirt, when Dr. Christian Friedrich Michaelis came calling in 1783.

3 | The Incident of Morgan's Bones

It may be asked why I [list] the mammoth as if it still existed? I ask in return, why I should omit it, as if it did not exist? Such is the economy of nature, that no instance can be produced of her having permitted any one race of her animals to become extinct; of her having formed any link in her great work so weak as to be broken. To add to this, the traditionary testimony of the Indians that this animal still exists in the northern and western parts of America, would be adding the light of a taper to that of the meridian sun.—Thomas Jefferson, 1781–82

THE prospect of a summer of adventure along the Conestoga wagon ruts to Fort Pitt and Kentucky's wild west beckoned Christian Michaelis. He literally itched to go.

On February 3, 1783, King George III gruffly proclaimed a cessation of hostilities between the armed forces of Great Britain and those of the United States of America. Two months later, General George Washington and Guy Carleton met at Tappan, on the fringe of the Westchester wilderness, to work out details of the departure of British troops from New York City and Long Island and the exodus of thousands of Loyalist families to Canada.

Under the subtle influence of the rum-and-brandy punch imbibed at these conferences, the essential morning exercise of the officers' horses, and the enforced sharing of one set of privies, the icy politeness that characterized relations between British and American staff officers and aides began to melt. Actual fraternization first occurred among the medical officers, who felt deeper loyalty to medicine and abetting the recuperative potentials of men than they did toward any political system. There were visits across the lines to examine hospital facilities and to discuss operating techniques, followed by evenings of toddy, whist, and wenching. Subsequent friendships developed between race-horse devotees, cardsharps, and natural philosophers.

By May, 1783, Dr. Johann David Schopf, of Wunseidel in the Thuringer Wald, had persuaded both his superiors and American officials into giving him permission to undertake a year-long tour of the new confederation, from Massachusetts to Georgia. (His book, *Travels in the Confederation,* would become a classic report about regional characteristics of America at the end of the Revolution.)

Christian Michaelis similarly campaigned for a trip to Philadelphia, Fort Pitt, and that salty marsh in the Ohio River valley said to abound in giant teeth and bones. The wardroom chats at New Windsor, the talks with the Reverend Annan, and subsequent discussions with both Continentals and Loyalists had convinced him that he must make the journey. There were, he had been assured, at least two collections of the Salt Lick teeth and bones in Philadelphia. As for this American Philosophical Society, it seemed to be the most distinguished organization of and for naturalists in the New World. The membership included such notables as General Washington, Thomas Jefferson, Dr. Franklin, Benjamin Rush, John Bartram and his son William, and scores of physicians who, like Michaelis himself, had professional curiosity about bones and other specimens that might contribute to mankind's knowledge about its own origins.

So, with letters of introduction, numerous passes smeared with wax seals, and a purse filled with Spanish silver, Michaelis took a patrol boat across the Hudson, held his temper in check while American sentries pawed through his portmanteau, then boarded the canvas-hooded stage wagon for Philadelphia.

The dream of the wagon ride across the mountains dissolved during his first visit with William Bartram. Hatred of "the British and their hirelings" was too rampant at Fort Pitt and beyond, through the Kentucky wilderness. The scalping knife and torture stake were routine weapons in the frontier's version of warfare. British agents from Niagara were being blamed for

the sporadic ambushes and kidnapings in the Ohio valley. Both
British and Indian belligerence precluded American hunting,
trading, and settling along the Great Lakes. The odds were that
no wagon driver would take the chance of carrying a British
officer across the mountains, pass or no pass.

The war's impact on the American Philosophical Society was
almost as discouraging. Since he was, by far, the most distinguished researcher and advocate of natural philosophy in
America, Benjamin Franklin was automatically re-elected
president of the Society. But he had been in France since the
fall of 1776. Many of the members had served with the Continental armies. The bones, rocks, stuffed animals, and other
objects in the Society's cabinet had been carted off to members'
homes during the British occupation in 1777–78. The sporadic
meetings since then were primarily devoted to discussions
among the physicians of operating techniques, rehabilitation of
the wounded, the possible causes of yellow fever, and similar
medical matters.

There was a small collection of unidentified teeth and bones
recently forwarded from Virginia by Thomas Jefferson. This,
Society members assured Michaelis, boded well for reorganization of the Society and restoration of its cabinet of natural
wonders. Jefferson was returning to Congress, after two tur-
bulent years as Virginia's Governor. He was an avid naturalist
and deeply interested in this mysterious beast that Dr. Franklin
called *Mammut americanus.* Mr. Jefferson agreed with most
of the clergy that the *Mammut* was not extinct and that herds of
the beast were still wandering the prairies or mountains of the
Far West, perhaps even in the unexplored forests of Kentucky
or Ohio. But research into these theories was still in the future.
Currently, the best collection of *Mammut* bones in Philadelphia
was owned by Dr. John Morgan.

John Morgan had used a few of the pieces in his classes, to
demonstrate cellular structure, dentine layering, and the like.
But he refused to sell any of them. He shared Dr. Franklin's
hunch that matters of huge significance to medicine, to naturalists, hence to all mankind would one day emerge from studies

of such obviously ancient objects. This was not, of course, a matter that one discussed publicly. Neither Jefferson nor General Washington nor Congress could be expected to endorse such radical research. They were politicians, hence were slaves of public will and, alas, public ignorance. Support for the research would have to come from people who just didn't give a damn about theology. Perhaps the year would come when he could close his office, get out that muddy box, assemble every scrap of evidence he had laid away since student days in Glasgow, and puzzle out the past, present, and future significance of the *Mammut* and similar creatures.

However, if Dr. Michaelis wanted to have accurate drawings of the Big Lick artifacts to take home to his father at Göttingen, that fellow Peale was worthy of the assignment. He was an excellent portrait painter, a zany inventor, and as fecund as a dairy bull.

The economic urgencies that forced Charles Willson Peale to welcome a Hessian physician to his home at Third and Lombard streets and accept the assignment to do concise drawings of Dr. Morgan's boxful of old bones echoed from the front door to the sunflower-screened outhouse. Infants, toddlers, dragon chasers, and rescuers of maidens fair bellowed, clanked, paced, daydreamed everywhere. Up and down the front hallway, with the intensity of Saint George on bony nag, a red-haired boy pedaled a three-wheeled vehicle through endless loops. (The vehicle, Peale explained, was his latest invention, a mobile toy he called a *velocipede;* if it proved popular, he might tinker with a more substantial model for adults.) From a front bedroom, the singsong voice of a small girl was imprecise counterpoint for the bellowing of an angry baby. A youth lying, belly flat, with an open book upon the parlor floor fixed unblinking bluebutton eyes on Michaelis. A medley of laughter and jeers echoed through the open window of the studio. A pantalooned youth with the same hawk nose, blue eyes, and long fingers as Peale absorbedly clicked the set of false teeth mounted on one of the worktables.

Drawings of old bones, sir? Drawings of almost anything that

would yield a few coins! Rachel Brewer Peale was lusty and
fertile too. There had been eight children so far; four had survived the nightmare months of infancy. And, judging from
Rachel's robust health, there would be more. He did have an
order from Benjamin West for drawings of American uniforms,
and there was talk of a triumphal arch to honor Dr. Franklin's
homecoming with the peace treaty. But . . . drawings of Dr.
Morgan's boxful of bones? A matter of mere days: where were
they?

At the time, neither Michaelis nor Peale realized the impact
this seemingly simple project would have on the fields of
natural history, paleontology, and applied science. There was
little reason, then, for either of the men to record the details of
their agreement, the length of Michaelis's visit to Philadelphia,
or the matters that preoccupied them before and after the discussions about the drawings. The only clue Christian Michaelis
provided to his Philadelphia visit was in a letter he wrote to
Göttingen from New York City on August 7, 1783. He made
some comments on medical techniques in America—undoubtedly influenced by conversations and classroom visits with John
Morgan and Benjamin Rush—and mentioned his agreement
with Peale to obtain mathematically precise drawings of Dr.
Morgan's *Mammut* bones.

In Charles Willson Peale, Michaelis had found the perfect
person to bring the wonders of natural history into full public
view and thereby touch off the widespread curiosity that would
lead to further discoveries, further research, and the beginnings
of paleontology as an accepted scientific study.

The son of a convicted forger who had been banished from
England to live in exile in Annapolis, Maryland, Peale discovered his artistic talents at an early age. After his father's
death, in 1750, nine-year-old Charles found solace and escape
from the dreary poverty of his surroundings by doodling pen-
and-ink sketches and copying illustrations from the few books
he had access to. Soon he was designing patterns for the needle-
work his mother did to support herself and her three young
sons.

But there was not much profitable work for a child artist in Annapolis, so, at thirteen, Charles was apprenticed to a saddle-maker. By the time he was twenty, a year before his apprentice-ship was officially completed, he had set up his own saddlery and was experimenting with watchmaking, jewelry design, sil-versmithing, and sign painting. In 1762, he married seventeen-year-old Rachel Brewer.

After his marriage, Peale began to devote more and more of his time to painting. His first efforts at portraiture won a con-siderable amount of praise, and within a year he was accepting commissions from wealthy Annapolis planters and merchants.

Despite this promising beginning, Peale could not earn enough to cover the debts he had been incurring in his saddle-making business, and he was forced into bankruptcy. Aided by his brother-in-law, he fled to Boston to escape his creditors. There he sought out John Singleton Copley, New England's most distinguished portrait painter, and persuaded him to ac-cept his services as a part-time assistant in exchange for lessons.

When Peale learned that Rachel had given birth to their first child, a son, he hurriedly left Boston. He got as far as New Jersey when he ran out of money; in order to earn enough to get back to Annapolis, he went from door to door offering to do family portraits for a small fee.

By the time he got back to Annapolis, Rachel's father had died, leaving her with an inheritance large enough to cover Peale's debts. With his creditors placated, Peale was back in the good graces of the community and before long was once again receiving commissions from prominent citizens.

In 1767, several of his wealthy clients decided to send Peale to London to study with the eminent artist Benjamin West. West took an instant liking to Peale and coached him not only in portraiture but also in the intricacies of miniatures, mezzotints, and sculpture. He also introduced him to his good friend Ben-jamin Franklin. This was the year of the controversy in the Royal Society over the Big Lick bones, and undoubtedly some of Franklin's overwhelming interest in the issue, and in natural philosophy in general, rubbed off on Peale.

Peale did well in England. He was elected to London's Society of Artists, and his work was displayed beside that of West, Copley, and other prominent painters in London galleries. When he returned to Annapolis in June, 1769, he was welcomed as Maryland's "first trained artist," and he received enough commissions to keep him busy for a full year.

As busy as his painting kept him, Peale managed to find time to indulge his inquisitiveness and pursue a host of developing interests. After working all day in his studio, he would spend his evenings conducting scientific experiments and tinkering with the inventions that grew out of those experiments. Among his myriad creations were a painter's quadrant with a telescopic eyepiece; an apple-paring machine; a "portable arm," based on a design by Franklin, that could be used to pick fruit from a tree or pluck a book off a high shelf; a spillproof milk pail; a portable steam bath; and the velocipede.

The paraphernalia of Peale's "extracurricular" pursuits eventually overflowed into his studio, where they became conversation pieces, helping to break down some of the social barriers between him and his sitters. One of these sitters, Martha Custis Washington, was so impressed by Peale's talents, his facile mind, and his wide-ranging knowledge that she invited him to meet her husband. Thus began a lifelong friendship that established Peale as the most famous portraitist of George Washington and led to his appointment as an official painter of battle scenes during the eight years of the Revolution. He was made a captain in 1776 and served at the defense of Philadelphia, at Valley Forge, and at the battles of Princeton, Trenton, and Yorktown. His family moved to Philadelphia after the British retreat in 1778.

By the time Michaelis and Peale met in 1783, Peale was once more thoroughly involved in his artistic career. His initial reaction to Michaelis's request was to regard it more as an interruption than anything else. He of course welcomed the opportunity to earn some extra money for his rapidly expanding family, but he was not particularly excited about the material he was being asked to draw. Dr. Morgan's mud-caked relics seemed identical to those Benjamin Franklin had been arguing about fifteen years earlier.

Within a week, however, the bones assumed new importance. Once word got around that Peale was doing the drawings, members of the Philosophical Society began dropping by to look at Dr. Morgan's specimens, "now that they've finally been cleaned." They were followed by students from the College of Medicine, who made a great fuss over the bones and begged for "any of the sketches that you don't send to that Hessian." Then came the ministers, deacons, and theologians eager to examine the remains of the "evil victims of the Flood" and discuss the dire implications of the relics. The stream of gawkers grew steadier and more persistent each day.

The drawings were finally finished and ready for shipment, but the bones were still arranged on Peale's worktable on the Sunday afternoon that Peale's brother-in-law and attorney, Colonel Nathaniel Ramsay, arrived from Baltimore. Ramsay had not heard about the bones, so Peale spent several hours telling him the story of Dr. Michaelis's quest and relating some of the many theories his visitors had offered about the origins of the remains.

Colonel Ramsay was fascinated. He picked up the bones, stroked them, turned them in his hands to view them from various angles. Finally he looked up at his brother-in-law.

"Charles," he said, "I wonder if you realize what you've got here. I would walk twenty miles to see this collection. Obviously, many others feel as I do. Why not, then, add these to your display of paintings? Obtain more of these oddities of nature—develop a museum and charge admission. It could not only become a source of income, but would it not also increase the sale of prints and miniatures?"

4 | A Museum of Natural Curiosities

When some of the first discovered bones of the Mammoth were eighteen years ago brought to you for the purpose of making drawings from them they were put in one corner of your picture gallery, where they fixed the astonishment of every visitor and daily served to confirm your intention of procuring, if possible, an entire skeleton. This your persevering zeal has at length accomplished; but a more extensive benefit was likewise the consequence; and the Museum, of which you are the founder, already rivalling many in Europe, is to be ascribed to the same cause, and dated from the same period.—Rembrandt Peale to his father, Charles Willson Peale, July 18, 1803

A Peale Museum! Colonel Ramsay's suggestion haunted the painter as he cleared studio worktables and laid out supplies for the rough drafts and mock-ups of the Triumphal Arch.

The need was so obvious—and so pressing—in this brand-new nation. Here were three, perhaps even four, million human beings, with thirteen individualistic governments now daring to experiment toward genuine confederation. The ideal was development of laws, of educational facilities, of commerce and industry that would enable a degree of human dignity never before achieved. But could such a dream be realized in a state of ignorance? And "state of ignorance" was an apt summation of everyday American life. Changing that would require much more than schools and books and teachers. The lessons for shaping life patterns and developing human dignity should not be limited to classroom recitations and sermons. Museums could become tremendously important teachers for *everybody*.

Look at these thirteen states. They were perched on a toehold of seashore, river plain, and piedmont between the Atlantic Ocean and the Allegheny Mountains. None of them was more than three hundred miles wide. The confederation and its awesome pledge to ensure for everyone the benefits of "life, liberty, and the pursuit of happiness" could endure *only* if its citizens swarmed across the Allegheny barrier into the forests and onto prairies of the West. The recent migrations into Kentucky and the agitation for land grants in the New York and Ohio wildernesses foreshadowed the "rush for the West" that must occur if the confederation was to survive.

Yet most Americans were completely ignorant about regions more than a day's jog from their homes. A Yankee sea captain might be familiar with the Portuguese and Dutch slave pens on the African coast and the plantations and bawdy houses of Havana, yet have no knowledge whatsoever of life in the Ohio woods. Similarly, the Pennsylvanian, though perhaps sophisticated in European affairs, knew next to nothing about the landscapes, wild animals, minerals, vegetation, or even the climatic peculiarities of the American vastness beyond his state boundaries.

A repository, then, that would display the creatures, plants, rocks of each region of America could become a citizens' university. There parents and children would learn, each at his or her own pace, the verities of nature and her savage brood in each region of the confederation—and in those lands of promise beyond the mountains.

Theater! There was a magic word. Consider the foul-smelling oil that could be stewed and skimmed from the carcass of a snake. Dreadful stuff. But put two ounces of it in a pretty bottle and glue on a label that promised a "miracle cure" for almost anything. Flaunt these pretty bottles on a street corner while you recite gibberish in stentorian tones, deftly interspersing your speech with a few sleight-of-hand tricks. Unitarians, Baptists, Quakers, even Amish paid a guinea a bottle for the stuff and blessed you for letting them have it. Theater!

The enthusiasm and eagerness with which people flocked to see Dr. Morgan's *Mammut* splinters demonstrated the same human urge. The bones had been in Philadelphia for fifteen years. Then, suddenly, because of a persistent Hessian physician, the element of theater took over, and the desire to *see* spread like swamp fever. The human attraction to spectacle, to theatrics, was a powerful force. It should be used to educate as well as to entertain. A museum was most assuredly a vehicle of education—an important learning place for adults as well as youngsters.

Memories of the British Museum in London still sent shivers through Peale. The place looked like a junk dealer's storeroom. Its displays were dusty mounds of old bones, rotten sticks, withered flowers, and faded manuscripts stacked in cases or on funereal black, tan, or gray tables. The identifying labels were in somber Latin, lettered in heavy India ink.

Any museum that would stimulate, hence open the mind for learning, must have a dash of theater. Its bobcat must not be a mangy hank of fur sprawled in display-case shadows but a sinuous predator with hypnotically evil eyes and gleaming fangs, seemingly intent on springing out of a tree branch and, with devilish screams, terrifying every person in the room. A rattlesnake must not be a faded diamond-patterned skin nailed on brown wood, but a coiled threat on a boulder, with fangs exposed and muscles swollen for that strike of death. These were the life forms that the frontiersmen would have to subdue in the land beyond the mountains. These were the theatrics that a museum would have to use to fulfill its role as a worthy medium of education. Morgan's *Mammut* fragments, like the animals, the flowers, and the mineral wonders of America, must be exhibited in natural settings and against background paintings that would give a sense of the topography and the surrounding plants, animals, and rocks.

Thus the Peale Museum—he smiled when he said the name to himself—could become not only a medium for education and a source of income but would also be a continual showplace for his skills as an artist, as well as a splendid training place for his sons.

Peale had named his three sons—Raphaelle, Rembrandt, and Rubens (a fourth, Titian, was born in 1799)—after the great masters, and it seemed as if they were already beginning to live up to their names. Raphaelle, ten, demonstrated a sensitivity for still lifes and portraits. Rembrandt, going on six, daubed rather good floral designs and likenesses of household pets on those cedar shakes and scraps of newsprint he was permitted to use for "drawing things." Rubens was still at the breast. But he had long, sensitive fingers and a way of staring at you so intently that you half expected him to gurgle a request for pencil and paper and scrawl a profile caricature.

Since the *Mammut* bones had proved their drawing power as a "great curiosity," Dr. Morgan must have been one of the first to respond to the dream of the Peale Museum. He was relieved to learn that Peale wished to keep the collection on display and thus spare the Morgan household the constant requests to examine "those bones from the Deluge."

But beyond a rearrangement of the Morgan collection in the front parlor, no efforts toward actually setting up the museum were undertaken before the spring of 1784. Then, a catastrophe that incapacitated Peale for weeks seems to have been a decisive factor in resurrecting and implementing the idea.

Throughout the summer and fall, the painter focused all of his energy on the triumphal arch that was to span Market Street. The six hundred pounds appropriated by the Assembly was enough for only a canvas-and-timber construction. Peale strove for a magnificence that would evoke demands for its recreation in limestone, marble, and bronze as an enduring memorial to the birth of the United States. More than sixty feet wide, with its central figure of Peace poised sixty-five feet above the street, the structure was to have three arches separated by Ionic columns and topped by a Romanesque balustrade. Framed in the balustrade would be a series of paintings by Peale of Indians building churches in the wilderness, the Liberty Tree with thirteen branches of fruit, General Washington as Cincinnatus returning to his plow, and various other symbols, including the lilies of France, a bust of Louis XVI, and garlands of Pennsylvania wildflowers.

The Assembly's Committee for Celebration of the Peace approved not only midwinter construction of the canvas-and-timber structure but also Peale's plan for illuminating the arch with 1,150 oil lamps and for setting off a fusillade of rockets and Roman candles at dusk on the day of its unveiling.

The framework of the arch was being notched and bolted together when General Washington rode through Philadelphia on December 8, en route to Annapolis to resign his commission as commander in chief of the Continental armies. Peale persuaded him to stay overnight so that he could pose for a "new likeness" to be incorporated into the gallery of paintings on the arch.

Congress approved the treaty of peace on January 14. The news reached Philadelphia that night; an emergency session of the City Council decided that Philadelphia's formal celebration would begin on the 22nd with a proclamation read at the court-house by the sheriff and the unveiling of the arch at dusk. No other lights would be permitted along Market Street that eve-ning so that the Arch might radiate its promise of peace and confederation without interference.

Sunset was spattering the shadows with rose and lavender when Peale tacked the last painting into place and began lighting the lamps. Carriages, gigs, and Conestogas had rocked in from Chestnut Hill, Valley Forge, and distant York. They massed along Market Street for blocks. On the platform be-neath him, a half-dozen artillerymen stood with flashpans at the ready to light the seven hundred rockets and Roman candles.

Peale was atop the balustrade lighting the last row of lamps when a fusilier suddenly reached down and fired one of the largest rockets. The squad's sergeant shouted, ran over, and stamped out the sputtering fuse. There was a gush of flame as the rocket soared above Market Street, igniting the wet varnish on the paintings and frames.

The artist's heart began to pound so savagely that the shouts from the crowd, the whinnies of the horses, and the hiss of the burning canvases seemed a dim and distant bedlam. Now the other rockets were firing. The artillery sergeant, screaming, beat at his flaming beard and jumped to the street.

Peale never could recall how he got to the ground. Later, neighbors told him that he had appeared on the rocket platform, kicked his way through the flames, and then leaped to the street. But the image of the artillery sergeant's charred body sprawled like a rag doll on the Market Street cobbles, the screams of the children and horses, and the hell of the arch aflame in the lavender dusk erased all other memories.

A carriage brought Peale back to Rachel an hour after the explosion. His coat and trousers were reduced to charred tatters. Blisters had started to form along the white welts of burns on his hands, face, and ankles, and two broken ribs stabbed him with each breath. His injuries kept him in bed for almost a month.

The morning after the fire, he dictated a message to the mayor and City Council offering to restore the arch and the paintings if they would agree to another formal proclamation of peace. The officials concurred but remained deaf to demands that Peale be awarded another appropriation for the reconstruction. To curb both his resentment of the Council's miserliness and his roars against his invalidism, Rachel, sputtering about an "indignity," lured him back to his plans for the museum.

The indignity she was referring to was the Pennsylvania Legislature's refusal to repeal a law forbidding theatrical performances in the Commonwealth. Now, it was said, the influence of the Methodists, Presbyterians, and Baptists was so great that an even sterner law would be introduced forbidding performances of pantomimes. This implied, Rachel pointed out, that a new form of entertainment, acceptable to the sanctimonious Scotch-Irish and Amish delegates, could become a most profitable undertaking.

The silence from the bedroom signaled Rachel's victory. Peale was plotting methods by which he could incorporate a maximum of theater into the Peale Museum, with a minimum of cries of "godlessness" from York, Lancaster, and Pittsburgh politicians.

Refinancing and restoring the arch occupied Peale for two months. The site was moved to Chestnut Street, near the Statehouse. The use of rockets, Roman candles, and other pyrotechnics was forbidden. On May 10, the proclamation of peace was formally reannounced, and the 1,150 lamps were safely lighted. Peale strode home through crowds of admirers to regale Rachel with the details.

Over the next three years, development of the museum became Peale's major interest. David Rittenhouse, recently appointed professor of astronomy at the University of Pennsylvania, doubted that a museum could ever be profitable enough to support a family, especially one as sizable as Peale's. Dr. Franklin and Dr. Rush were almost as cautious. William Bartram was enthusiastic and eager to co-operate. During the 1760's, his father had urged Dr. Franklin to seek a royal grant for an expedition to study the botany, zoology, soil types, and rocks of all British North America and, if possible, to determine a feasible overland route to the Pacific, north of New Spain. William Bartram possessed the same zeal for knowledge about the West, and was appalled by the ignorance of the war veterans agitating for "land grants beyond the Alleghenies." A museum devoted to natural history, he believed, was as vital to the public good as a university. It would supplement, rather than compete with, the botanical gardens his father had established at Kingessing on the Schuylkill River. William himself was working on a series of drawings of 215 species of American birds, and he wished to have many of the specimens stuffed and mounted for display at the botanical gardens. He offered to teach Peale and his sons Raphaelle and Rembrandt the techniques of taxidermy.

The apprenticeship began during the summer of 1784. Experimenting with skins Bartram supplied, as well as with the carcass of Dr. Franklin's Angora cat, Peale learned that the essential factor in "imparting natural grace" to a stuffed animal was to carve a skeleton in wood that would "impart the proper swellings of the muscles." The arsenic-and-alum-cured skin was stretched over the wood skeleton.

This dedication to realistic detail was typical of the patience and imagination Peale applied in creating exhibits for the museum, and accounted for the sparsity of the exhibits during 1784–86. Experiments in three-dimensional backgrounds, radically new types of lamps, ventilating systems, and the first American appearance of the "moving picture" were all explored and audience-tested before the initial advertisement for Peale's Museum was published in *The Pennsylvania Packet* of July 18, 1786.

In the spring of 1784, Peale learned about the Eidophusikon, a device invented in London in 1780. It consisted of a sequence of paintings on glass, which, when reflected on a screen, appeared to be moving. Introduction of the Eidophusikon to Philadelphia, Peale reasoned, would not only attract enough paying viewers to finance the construction of the galleries and auditoriums needed for the museum, but would also provide daily opportunities to study audience response and perfect that essential ingredient the Peale Museum must have: *showmanship*.

The premier performance of the Eidophusikon on May 20, 1785, was advertised as "Perspective Views of Changeable Effects, or Nature Delineated and in Motion." The five scenes on the program depicted a bucolic sunrise, night and dawn in the city, the "city and proud seat of Lucifer" described by Milton in *Paradise Lost*, a thunderstorm over a Roman villa, and Vandering's Mill on the Schuylkill.

But Rembrandt, his playmates, and other children shushed into the auditorium by culture-conscious relatives refused to stumble through pronunciations of either "Eidophusikon" or "Changeable Effects." With the simple directness characteristic of children, they began referring to the show as "moving pictures." Peale recognized the superior showmanship of their phrase. By the time he had added a sequence on "The Battle of the Serapis and Bon Homme Richard" to the program in the winter of 1785–86, he was promoting the show as "The Moving Pictures." His advertising is believed to be the first American appearance of the term.

Showmanship that antedated the slickness of P. T. Barnum by a lifetime was displayed in the corridor leading to the "Moving Pictures" auditorium. Both walls were hung with battle scenes, portraits, and groups of miniatures painted by

Peale and his brothers. Mounted on shelving beneath them were the experimental groupings for the museum—Dr. Morgan's *Mammut* bones, a rare paddlefish from the Allegheny River, and stuffed birds and ducks labeled "Do Not Touch. These Are Covered with Arsenic Poison."

Two other inventions appeared in the auditorium during 1785. One was a system of propellorlike fans suspended from the ceiling and powered by a boy turning a crank. Thereafter "fresh, cooling air" was a standard attraction of the museum. There were also oil lamps with metal and glass reflectors; Benjamin Franklin brought the design home from France and boasted that "Each provides the light of two hundred candles and in burning consumes its own smoke." Both of these gimmicks lured customers and helped establish Peale's reputation as a master showman.

But ticket sales ebbed during the winter of 1785–86. Most Philadelphians had seen the moving pictures at least once. The Peales were too busy with portraits, miniatures, and construction of three-dimensional displays to think about changing the program. In the spring of 1786, Peale decided to abandon the moving pictures and concentrate on the museum.

The first announcement that "Mr. Peale will make a part of his House a Repository for Natural Curiosities" appeared in *The Pennsylvania Packet* of July 18, 1786. Three days later, Peale was elected to the American Philosophical Society and began to make requests for fossils, animal skins, dried flowers, minerals, and other "oddities of Nature" from correspondents of the Society in Virginia, the Carolinas, Kentucky, and western New York.

On September 14, the Annapolis Convention, an assembly called by five states to discuss interstate commerce, resolved to invite delegates from all States to Philadelphia during May, 1787, to "devise provisions necessary to render the Constitution of the Federal Government adequate to the exigencies of the Union." The Constitutional Convention would bring important politicians and the wealthiest landowners and merchants from each State to Philadelphia.

All through the following winter and spring, the Peales, coached by Franklin, Rush, and the Philosophers, worked diligently to complete the museum exhibits. The results of their efforts were described by the Reverend Manasseh Cutler, a scientist and one of the colonizers of the land along the Ohio River, who visited the museum on July 13, 1787.

"Immediately after dinner, we called on Mr. Peale," the Reverend Cutler recalled, "to see his collection of paintings and natural curiosities. We were conducted into the room by a boy, who told us that Mr. Peale would wait on us in a moment or two. He desired us, however, to walk into the room where the curiosities were, and showed us a long narrow entry which led into the room. I observed, through a glass at my right hand, a gentleman close to me, standing with a pencil in one hand and a small sheet of ivory in the other, and eyes directed to the opposite side of the room, as though he was taking some object on this ivory sheet. Dr. Clarkson (my companion) did not see his man until he stepped into the room, but instantly turned about and came back to say 'Mr. Peale is very busy, making a picture of something with his pencil. We will step back into the other room and wait till he is at leisure.'

"We returned to the entry, but as we entered the room we came from, we met Mr. Peale coming to us. The Doctor started back in astonishment and cried out, 'Mr. Peale, how is it possible you should get out of the other room to meet us here?'

"Mr. Peale smiled. 'I have not been in the other room,' says he, 'for some time.'

" 'No!' says Clarkson. 'Did I not see you there this moment, with your pencil and ivory?'

" 'Why, do you think you did?' asks Peale.

" 'Do I think I did? Yes,' says the Doctor. 'I saw you there if ever I saw you in my life.'

" 'Well,' says Peale, 'let us go and see.'

"When we returned, we found the man standing as before. My astonishment was now equal to that of Dr. Clarkson . . . I beheld two men, so perfectly alike that I could not discern the minutest difference. One of them indeed had no motion; but he appeared to me to be absolutely alive as the other, and I could hardly help wondering that he did not smile or take part in the conversation. This was a piece of wax work which Mr. Peale had just finished, in which he had taken himself.

"This (entry) room is constructed in a very singular manner, for the purpose of the exhibitions, where various scenery in paintings is exhibited in a manner that has a most astonishing effect. It is very long but not very wide, has no windows nor floor over it, but is open up to the roof, which is two or three stories, and from above the light is admitted in greater or less quantities at pleasure.

"The walls of the room are covered with paintings, both portrait and historic. One particular part is assigned to the portraits of the principal American characters who appeared on the stage during the late revolution, either in the councils or armies of their country. The drapery was excellent, and the likenesses of all of whom I had any personal knowledge were well taken. I fancied myself introduced to all the General Officers that had been in the field during the war, whether dead or alive, for I think he had every one, and to most of the members of Congress and other distinguished characters. To grace his collection, he had a number of the most distinguished Clergymen in the middle and southern States who had, in some way or other, been active in the revolution.

"In other parts of the room were a number of fine historic pieces, executed in a masterly manner. At the upper end of the room, General Washington, at full length and nearly as large as life, was placed as President of this large and martial assembly.

"At the opposite end, under a small gallery, his natural curiosities were arranged in a most romantic and amusing manner. There was a mound of earth, considerably raised and covered with green turf, from which a number of trees ascended and branched out in different directions. On the declivity of this mound was a small thicket, and just below it an artificial pond; on the other side a number of large and small rocks of different kinds, collected from different parts of the world and represented in the rude state in which they are generally found. At

the foot of the mound were holes dug and earth thrown up, to
show the different kinds of clay, ochre, coal, marl, etc. which he
had collected from different parts; also various ores and
minerals.

"Around the pond was a beach, on which was exhibited an
assortment of shells of different kinds, turtles, frogs, toads,
lizards, water snakes. In the pond was a collection of fish with
their skins stuffed, water fowls such as different species of geese,
ducks, cranes, herons; all having the appearance of life, for
their skins were admirably preserved.

"On the mound were those birds which commonly walk the
ground, as the partridge, quail, heath-hen; also different kinds
of wild animals;—bear, deer, leopard, tiger, wildcat, fox, rac-
coon, rabbit, squirrel. In the thicket and among the rocks were
land snakes, rattle snakes of an enormous size, black, glass,
striped and a number of other snakes. The boughs of the trees
were loaded with birds, some of almost every species in
America, and many exotics. In short, it is not in my power to
give a particular account of the numerous species of fossils and
animals, but only their general arrangement. What heightened
the view of this singular collection was that they were all real,
either their substance or their skins finely preserved.

"Mr. Peale was very complaisant, and gave us every information we desired. He requested me to favor him with any of
the animals and fossils from this part of America, not already
in his museum, which it might be in my power to collect."

Cutler's impressions were particularly significant, for he was
to play a key role in furthering the development of the museum
in America and in initiating botanical and geological research
in regions that were just being opened to settlers.

On July 13, the same day that Cutler visited the Peale
Museum, Congress approved the Northwest Ordinance, outlin-
ing a governmental framework for the "Territory Northwest of
the River Ohio." Cutler, as one of the leaders of the Ohio
Company, had lobbied for and helped to draft the ordinance
and had convinced several New York bankers to finance down
payments on 6,500,000 acres of Ohio forest and river bottom

where land-poor, tax-ridden New Englanders could migrate. The Ohio Company's contract for land purchases from Congress's Treasury Board was signed the following October 27; soon wagon trains bearing the slogan "Ohio or Bust" began rumbling out of Connecticut and Massachusetts to make the long trip across the Alleghenies.

Cutler was also a botanist, and he was completing a work giving the first systematic account of New England's flora; he knew that a strange new galaxy of natural wonders would confront the Yankee immigrants in Ohio. New Englanders were mainly of British descent and were planning to follow the British tradition of developing diversified farms. This meant that most of the Ohio forests would be cut down. What flora and fauna would adapt to the new grasslands? What minerals would be found in the rocks and soils? Cutler, recognizing that the future of the new territory depended on the settlers' familiarity with nature and their ability to cope with their environment, was convinced that knowledge of the earth's secrets must become a cornerstone of American education. Museums like Peale's seemed a far more effective vehicle for such education than the dull rote learning, enforced by knuckle-raps, that was characteristic of traditional schoolrooms.

Cutler's work with the Ohio Company took him to Hartford for conferences with Israel Putnam and other organizers of the first "Ohio or Bust" wagon train. New Haven was a stagecoach stop on the way. Since Cutler was a Yale alumnus, it is more than likely that he met with Yale's president, the Reverend Ezra Stiles, to discuss Peale's Museum of Natural Curiosities.

The shrewdest and most pragmatic educator in New England, Ezra Stiles had been a member of the American Philosophical Society since 1768 and was an old friend of Benjamin Franklin's. Like Franklin and Cutler, he questioned the veracity of Christianity's dogma about recent Creation. There was, he agreed, dire need for the introduction of studies about minerals, soils, flora, and fauna in American universities. How could the trustees be convinced? Who was the right person to pioneer such courses?

While Peale was launching his museum and Cutler was helping to open the Ohio territory and seeking ways to educate the public about the natural sciences, Dr. Morgan's *Mammut* bones were making an important journey across the Atlantic Ocean.

In 1785, after visiting the British Museum and comparing the *Mammut* bones there with the Peale drawings, Christian Michaelis had written an article for a German magazine about *Mammut americanus*. The bones of the Morgan collection, he concluded, like those given him by the Reverend Annan, came from an animal that possessed neither tusks nor trunk and was probably extinct.

Soon after the article was published, Petrus Camper, professor of medicine, surgery, anatomy, and botany at the University of Groningen in Holland, wrote to Michaelis asking if he might examine the Peale drawings. A few months later, Professor Camper asked Michaelis to negotiate with Dr. Morgan and Peale for the sale of the bones.

Dr. Morgan rejected the offer in 1785, but in 1787 accepted a second offer of a hundred guineas from Professor Camper.

Thus, about the time of the Reverend Cutler's visit, Peale and his sons tenderly wrapped the yellowed teeth and bone fragments in soft cloth and burlap, packed them into a stout wooden chest, and sent them down to a Delaware wharf. It was, for Peale, emotionally comparable to a death in the family. The bones had been the magic charms from which the reality of his Museum had slowly materialized. Now the barren spot they had occupied on the display tables haunted him. What was this strange giant the bones came from? Were its descendants still trumpeting in the Ohio wilderness? Or, as Ben Franklin had intimated and Dr. Michaelis now claimed, was it extinct and thus, by implication, a challenge to the validity of the recent-Creation theory? Somewhere, somehow, the Peale Museum must obtain a complete *Mammut!*

Thus, in its own devious ways, past influenced future. In Europe, the Morgan bones would touch off bitter philosophical debates about the Earth's age and the massive forces that determined the order and relative position of the strata of the Earth's crust. At Yale, Harvard, Columbia, and Princeton Universities, museums in imitation of Peale's would be set up. And, in the terraced valley of the Wallkill, drainage ditches for new wheat and corn lands would be dug closer and closer to strange, moss-covered mounds in the swamps.

5 | Ninth Wonder of
the World

We are forced to submit to concurring facts as The Voice of God. The bones exist. The animals do not.—Rembrandt Peale, 1803

BETWEEN 1790 and 1803, a few "radicals" enabled the Peale Museum to become the "Ninth Wonder of the World" and provided theories that would justify educational pioneering in three new fields: geology, engineering, and paleontology.

One of the most important figures in this movement was Dr. Caspar Wistar, a veritable Daniel Boone of science now recalled mainly as the person for whom the genus of the flowering vine called *Wistaria* was named. Grandson and namesake of a New Jersey glassmaker, Wistar was fourteen years old when the Revolution began. After serving as a volunteer nurse and orderly and then as a surgeon's assistant at Valley Forge, he decided to pursue a career in medicine; when the war was over he enrolled in the College of Medicine in Philadelphia, where he studied with Dr. Morgan. Wistar did so well in Morgan's classes that he was awarded a grant for graduate study in Great Britain. He sailed to London in 1784, studied there for a year, then went north to the University of Edinburgh, where he earned his M.D.

A few of the lecturers at Edinburgh agreed with the radical theory of *Uniformitarianism,* advanced by the geologist James Hutton. An inveterate wanderer, constantly recording rock colors and contours in a smudged notebook, Hutton had concluded that the interlacing of limestone, schist, and red granite in Scotland's glens and loch basins had been created by unknown forces that were continually building, destroying, and rebuilding the Earth's surface. One of these forces, he was certain, was volcanic action. Another was earthquakes. A third must be the persistent "gnaw" of wind and rain, freezing and thawing.

Hutton had no intention of attacking Christianity or its clergy. He was a devout believer in an Almighty God. But he could not accept the implication that his God was whimsical enough to have created the Earth 6,000 years ago and then, like a mad baker, sprinkled obsidian daggers, seashells, giant bones, and curlicued layers of rock into the batter. God, to James Hutton, was Awesome Majesty; His time was ancient beyond comprehension; His subtleties were beyond the ken of mankind, as they should be.

Arguments about "old Hutton's daydream" stimulated Wistar's interest in the science recently named *geology* (earth science). And, since he specialized in anatomy, the age and contours of mysterious bonelike objects intrigued him. In 1787, a year after he received his M.D. degree, Wistar applied for membership in the American Philosophical Society. He soon became involved in analysis of what was probably the first dinosaur bone ever discovered in North America. The coherence of his reports won the respect of Thomas Jefferson, an avid collector of fossils; correspondence between the two men grew into a lifelong association between Jefferson the collector and Wistar the analyst.

The constancy of this relationship is somewhat puzzling, since Jefferson had publicly expressed the belief that "such is the economy of nature, that no instance can be produced of her having permitted any one race of her animals to become extinct." Dr. Wistar was convinced of the stupidity of the belief in recent Creation and certainly argued its implications with Jefferson. In private, Jefferson may have agreed. But in public, he was forced to be a "servant of the people," dependent on the votes of farmers and frontiersmen. Thus Jefferson's "naïveté" about geology and natural history was probably as studied as some of his successors' attitudes toward "balancing the budget" and "maintaining the peace." Open acceptance of such "devil's tools" as uniformitarianism or an Earth-before-Eden theory would have been a political liability.

A professor of chemistry as well as a lecturer on anatomy before he was thirty, Dr. Wistar was elected curator of the American Philosophical Society in 1793. He spearheaded a movement to combine the Peale Museum with the Society's cabinet of fossils and other "natural curiosities." Negotiations dragged on for several years, but finally Peale and the directors of the Society reached agreement. The Peale Museum was incorporated as a "public institution," and all of its displays were moved three blocks to the Society's brick-and-marble home adjoining Independence Hall.

Thomas Jefferson, enthusiastically serving as president of the Society, accepted the presidency of the museum's first board of directors and not only persuaded most of the Republican politicians in Washington to buy annual passes, but also routinely sent boxes of "strange bones" and other natural rarities to Philadelphia for Dr. Wistar's decision as to their worthiness as museum displays. "A specimen of each of the several species of bones now to be found," he wrote, "is to me the most desirable object in natural history."

By June, 1801, the museum seemed to be fulfilling Peale's goal of providing "a commendable source of education for everyone." Peale himself was redrafting a petition to the state legislature to establish a Pennsylvania Academy of Art, where "young gentlemen of promise might cultivate their skills." With classes ended for the term, Dr. Wistar was preparing to take his family to the seashore for the summer. But a letter from a Dr. George Graham at Montgomery, New York, caused him to postpone his own holiday. He sent his family off without him, promising to join them "as soon as we can raise some funds for Peale."

When he strode into Peale's studio a day or two later, Wistar was grinning broadly. Would the Peales, he asked, be interested in an expedition to New York's Wallkill valley, with a drawing account of five hundred dollars to be supplied by the Society? The purpose of the expedition? How about a complete *Mammut* skeleton?

Before the stunned Peale had a chance to find his voice, Wistar read him Dr. Graham's letter.

Two summers before, Graham had written, a farmer named John Masten began trenching his peat bog on the east bank Wallkill shoreline. Masten's log house, lean-to barn, and hog yard were four miles downstream from the Reverend Annan's farm, where eighteen years earlier Christian Michaelis had searched for *Mammut* artifacts. A few yards from the shore, the ditching crew excavated what appeared to be a hardwood log, four feet long, eighteen inches in circumference, but shiny black. The grooved spherical knob at one end caused the foreman to send a boy back to the barn for a crosscut saw.

Beneath the black patina, the "log" was clean-grained bone. It was obviously a thighbone, six or seven times as large as the thighbone of a stallion. It must be, the foreman guessed, another one of those "giants from Noah's Flood" that Preacher Annan used to harangue about. The men began to tear into the peat again. By the time a boy had fetched John Masten in from barn chores, several pieces of rib as thick as the boy's wrist had been whacked up.

John Masten was a farmer. Each summer hour was precious. He ordered everybody back to the trenching. If the weather held, they could start harvesting the oats in the morning. Bones, he grumped, were fit only for the manure pile or to be ground into meal.

But tales about "the big bone from Noah's time" ran from farm to farm. A preacher and then a doctor rode out to Masten's, brooded on what they saw, and convinced him that a digging party might yield him enough skeletal parts to sell to a showman or perhaps to that new museum at Columbia College in New York City.

One weekend that fall, between the wheat harvest and hog-killing time, Masten hauled kegs of hard cider out to the bog edge and invited neighbors in for a digging party. But neither

he nor the preacher nor the doctor felt it worth while to urge
caution in removing any bones that might be unearthed or to
sketch any oddly shaped or varicolored patches of dirt. So the
excavation became a kind of free-for-all in which most of the
bones, tusk parts, and teeth were broken. Then, Rembrandt
Peale reported years later, "so great a quantity of water, from
copious springs bursting from the bottom, rose upon the men
that it required several scores of hands to lade it out with all the
milk pails, buckets and bowls they could collect. . . . They
even made and sunk a large coffer dam, and within it found
many valuable small bones. But so much water had risen in the
pit that they had not the courage to attack it again."

Although hundreds of visitors tramped a path to Masten's
ditch to gawk at the bones (and pocket souvenirs), no newspaper editor deemed the dig worthy of a story, and no showman
or devotee of natural philosophy made Masten an offer. But,
hopeful of some sort of a bid during 1800, Masten decided to
get the pieces under cover, and so carted them up to his barn.

They were still there in the spring of 1801 when Dr. Graham
read an article about the American Philosophical Society and
the Peale Museum and wrote Dr. Wistar. The promptness of
Dr. Wistar's response so pleased Graham that he offered the use
of his home as base of operations for the Peales, who planned to
arrive in Montgomery in early July.

Charles, Rembrandt, and Raphaelle Peale reached Montgomery on schedule and were driven over to Masten's farm by
Dr. Graham. Masten was busy with harvesting; his main concern was that there should be no traffic across his land or digging
in the pit before all of his small-grain crops were harvested.

A price of one hundred dollars "in silver or gold" was fixed
for the digging rights and the bones in the barn. Then, so long
as they kept out of the way of the harvest crew, the Peales were
permitted to examine and sketch the bones in the barn, wrap
them up, and send them by wagon back to Philadelphia.
Raphaelle went south with the bones, thoroughly briefed on the
report to be delivered to Dr. Wistar. Meanwhile, Charles Peale
crossed the mountains to Newburgh and took a packetboat to

New York City to rent a ship's pump, pipes, and other tools
needed to drain the pit.

"It was resolved," Rembrandt later explained, "to throw the
water into a natural basin, about sixty feet distant. An inge-
nious millwright constructed the machinery and, after a week
of close labour, completed a large scaffolding and a wheel
twenty feet in diameter, wide enough for three or four men to
walk abreast. A rope around this turned a small spindle, which
worked a chain of buckets regulated by a floating cylinder. The
water thus raised was emptied into a trough, which conveyed it
to the basin. A ship's pump . . . and toward the latter part of
the operation, a pair of half-barrels . . . assisted in removing
the mud. This machine worked so powerfully that in the second
day the water was lowered so much as to enable them to dig,
and in a few hours they were rewarded with several small
bones."

Raphaelle Peale returned to the dig with an urgent request
from Dr. Wistar to concentrate on looking for parts of a lower
jawbone. Masten's barn-floor pile had included enough leg,
backbone, and rib pieces to permit the reconstruction of two or
perhaps three semiskeletons. But there was no evidence whatever of a lower jawbone. Certainly the beast must have had
one.

The pumping and digging continued until mid-September,
with a crew of twenty-five "at high wages." A section of lower
jaw was uncovered but was "broken to pieces in the first
attempt to get out the bones and nothing but the teeth and a few
fragments of it were now found." The Society's five hundred
dollars and most of Peale's reserve cash were spent. At Dr.
Graham's insistence, a dig was undertaken on the farm of a
Captain J. Barber, eleven miles up valley. It yielded two rotten
tusks and some ribs—but no skull sections. "With empty
pockets, low spirits and languid workmen, we were about to
quit the morass," Rembrandt recalled. Dr. Graham pleaded for
"one final try," on the hill farm of a Peter Millspaw, west of the
falls of the Wallkill. A few bones had been discovered there in
1795. It was, said Rembrandt, a drowned forest "where every

step was taken on rotten timber . . . and a breathless silence had taken her reign amidst unhealthy fogs." Charles Peale agreed to gamble the last of his funds by hauling over the ship's pump and building another set of drainage troughs.

But two more weeks of trenching and digging yielded only a few ribs and some toe and leg bones. The scarlet-and-gold dazzle on the hardwoods heralded a ground freeze-up and snow. Anyway, they were just about out of funds—they had enough for only one more day of digging. Charles and Rembrandt spent hours deliberating. Finally the father shrugged, ordered the pit abandoned, and sent the crew to excavate a moss-covered mound ten feet away.

Three feet down, they unearthed a complete lower jaw in excellent condition, a badly decomposed lower jaw, and a skull. "The unconscious woods echoed with repeated huzzas," Rembrandt reported. "They could not have been more animated if every tree had participated in the Joy. 'Gracious God, what a jaw! How many animals have been crushed beneath it!' was the exclamation of all. A fresh supply of grog went round."

The Peales and the canvas-wrapped bones reached Philadelphia in early November. Assembly of the skeletons began the week after Thanksgiving. "The three sets of bones were kept distinct," Rembrandt wrote. "With the two collections which were most numerous, it was intended to form two skeletons." Analysis of the backbone and leg parts convinced Dr. Wistar that the *Mammut* had nineteen ribs. The skull sections showed supporting structures for both an elephantine trunk and tusks. But the tusks were two, possibly three feet longer than those of a modern elephant, and curved so oddly in toward the front legs that it seemed as though the creature must have constantly banged its shins on them.

So much more must be learned about the *Mammut*, Wistar decided, that it would be more honest to mount the skeletons without any tusks and not to make any guesses about the coloration and texture of its hide. The bones were glued together, then strung on wire. Facsimiles of missing ribs, tail joints, and jaw pieces were carved by Charles and Rembrandt Peale.

The popularity of the Peale Museum had caused some of the more introverted members of the Philosophical Society to grumble about the groups of giggling, whispering children who trooped through the building. Partly in deference to these members, and partly because of the more commodious quarters available, many of the Museum exhibits were moved across the street to Independence Hall in the spring of 1802. But the display of the first *Mammut* skeleton ever assembled was a milestone not to be denied installation at Society headquarters.

The main hall was stripped for the occasion. But what setting should they give the skeleton? Peale had pioneered in the use of environmental backdrops for other flora and fauna. Could the same technique be dared for the *Mammut?* Natural philosophers were in violent disagreement about the plant life and even the contours of the Wallkill valley when the great beasts lived there. Were the terraced cliffs evidence that it had been a seashore? Or, as the black loam and bogs suggested, had it been a tropical jungle? And what other animals and plants and crawling things had flourished there?

The expedient course was to dodge the issue. Rembrandt had drawn hundreds of sketches during the digs at Masten's, Barber's, and Millspaw's. From these he created a massive oil painting of the water wheel and pit at Masten's, with allegorical portraits of his father and the entire Peale family posed on the pit's bank beside one of the giant leg bones. This painting served as the backdrop of the exhibit. The tuskless carcass of *Mammut americanus* stood on a pedestal in the center of the hall, eerily lighted by Benjamin Franklin's French chandeliers. Ceiling-to-floor drapes of blood-red velvet enhanced the sense of mystery.

Advance showings for members of the Society were held during May, 1802. The guests included professors from Harvard, Yale, Columbia, and Princeton colleges, and newspaper editors from Boston, New York, Baltimore, and Charleston. A few days later, a more exclusive showing was held for President Jefferson and his cabinet. With his secretary, Meriwether Lewis, beside him, taking notes, Jefferson slowly circled the skeleton of *Mammut americanus,* carefully examining and commenting on it. The cabinet members hedged them so closely that all the eager newsmen could hear were some mutterings about "those jungles in Florida."

The reporters hurried back to the inns to dash off articles that would be sent to their newspapers on the morning coaches. Deadline pressures caused several to succumb to cliché and hail *Mammut americanus* as the "Ninth Wonder of the World."

In view of what was soon to happen, the phrase was neither trite nor exaggerated. Soon a young law student named Benjamin Silliman would cross the Yale campus to speak to the university's president, Timothy Dwight. At Judge Hoffman's office in New York City, Amos Eaton, his true career still unknown to him, crammed for the state bar examination. Thomas Jefferson and Meriwether Lewis planned the research necessary for a proposed expedition to Florida and Louisiana. In Great Britain, William Smith journeyed from Bath to Wales, examining rock strata and meticulously recording and sketching the types of fossil shells he found in each layer. At the Jardin des Plantes in Paris, Georges Cuvier labored on his massive manuscript, *Leçons d'anatomie comparée.* This was the beginning of a new era, an age of scientific awareness. The Peale Museum and the skeleton of *Mammut americanus* were its harbingers.

6 | Reasonable Doubts

By our own act of assembly of 1705 . . . if a person brought up in the Christian religion denies the being of God, or the Trinity, or asserts there are more gods than one, or denies the Christian religion to be true, or the scriptures to be of divine authority, he is punishable on the first offence by incapacity to hold office or employment ecclesiastical, civil or military; on the second by disability to sue, to take any gift or legacy, to be guardian, executor or administrator and by three years imprisonment without bail.—Thomas Jefferson, *Notes on Virginia*, 1781–82

PHILADELPHIA became the Athens of the United States. Exhortations about the "New Greece" helped spread the popularity of Greek Revival architecture, and towns along the Atlantic seaboard and deep into the Appalachian wilderness were given such classical names as Troy, Athens, Corinth, and Ithaca. Arguments with Great Britain kept the Great Lakes throughway closed, so post roads were built across the mountains to Pittsburgh, making Philadelphia the major port of trade to and from the Ohio valley and the Northwest Territory. Merchants and bankers sought better schemes for wresting "a good living and ten per cent" from the farm-based economy, while manufacturers, craftsmen, physicians, and scientists researched those "principles of Nature" that might help them broaden their services and profits.

The new research was encouraged by the Quakers and the Unitarians and quietly condoned by the Episcopalians (who, after all, were still struggling to erase the stigma of their Church of England heritage and Loyalist ties during the Revolution). Thus Dr. Caspar Wistar and other members of the American Philosophical Society felt free to question the orthodox Judeo-Christian interpretations of Genesis, to state that *Mammut americanus* was probably only one of a vast array of extinct animal and plant species, and to pioneer investigation of prehistoric life, rock varieties, and the "chemical mysteries of Nature."

But, although intellectual freedom flourished in and around Philadelphia, scientific inquiry was still considered "unChristian" in most of the United States. An orthodox Christian clergy dominated both mores and education as rigorously as had their predecessors in medieval Europe, and any investigations of old bones, botanic fossils, and rock strata were suspect. The sermons and church-school teachings of Congregationalists, Methodists, Presbyterians, Baptists, Lutherans, and the few Roman Catholics adhered to the belief that the date of Creation was irrevocably fixed at 4004 B.C.

In the South and West, most preachers were farmers or craftsmen who had "met the Spirit of God" during a camp meeting. The camp meeting was originated by Methodist preachers in Kentucky in the 1790's; it was soon picked up by other evangelical sects such as the Baptists and the Campbellites, and its popularity spread rapidly throughout the United States. It had strong social appeal: the communal feeling it created brought families closer together and encouraged new friendships, and the frenzied fire-and-brimstone sermons followed by wild stomping and chanting and impassioned, on-the-spot conversions provided a tremendous emotional release.

Yale alumnus Eli Whitney was indirectly responsible for the spread of the camp meeting. Soon after he became a tutor on a South Carolina plantation in 1793, Whitney invented an "engine," operated by a hand crank, that separated the tenacious seeds from the fibers of cotton bolls. It was so simple a device that any blacksmith could copy it; many did—the courts failed to enforce Whitney's patent.

The introduction of the cotton gin launched a trans-Appalachian race to transform the Alabama, Mississippi, and East Texas wilderness into cotton plantations. Consequently, the Southerner and his preacher-dominated culture dominated the lands west of the Mississippi until the 1850's. The camp meeting and the devout, fundamentalist belief in "the divine authority of the Scriptures" became the accepted way of life south of the Potomac and Ohio rivers and west to the Rockies. As a result, many fundamentalists were elected to state legislatures and to Congress; as a political bloc piously dedicated to the "old-time religion" they were powerful enough to curb scientific research throughout the nineteenth century.

In New England, too, most Congregational clergymen agreed that anyone who questioned the Judeo-Christian interpretation of Genesis was "a defiler of God's Word," and thus a heretic. Their social influence was even more weighty than that of the circuit riders and camp-meeting orators. "The power of the Congregational clergy was equivalent to a police authority," Henry Adams wrote in his *History of the United States During the Administration of Jefferson and Madison.* "Did an individual deny their authority, the minister put his three-cornered hat on his head, took his silver-topped cane in hand, and walked down the village street, knocking at one door and another of his best parishioners, to warn them that a spirit of license and French infidelity was abroad, which could be repressed only by a strenuous and combined effort. Any man placed under this ban fared badly if he afterward came before a bench of magistrates. . . . Excepting the medical profession, which represented nearly all scientific activity, hardly a man in Boston got his living by science or art."

So when, during spring plowing in 1800, Pliny Moody discovered tracks a foot long and "three toed like a bird's" in a red sandstone ridge on his Connecticut valley farm, he knew whom he must consult. "These are, without question," his Congregational pastor concluded, "the imprint of Noah's raven. It rested on that ledge and probably slept there before resuming the dangerous journey back to the Ark." Natural philosophers at Harvard College upheld this explanation. Years later, scientists identified this "imprint of Noah's raven" as the first dinosaur footprint ever found in the United States. In 1818, when dinosaur bones were discovered in similar red sandstone a few miles south of Moody's farm, clergymen maintained that they were "probably the remains of giant humans."

However, even clergymen were forced to recognize that as

the country expanded physically it also had to expand economically. A growing population and inauguration of large-scale
trade with the states and territories west of the Alleghenies
created greater and greater demands for industrial and agricultural products. Silica, limestone, and mica were all rocks, and
each was as essential to industry as the rocks pocked with gold
and silver. Lime and potash and, some claimed, even iron were
essential for successful grain and fruit harvests and healthy
livestock. The need for more spinning mills to process the
South's vast new crops of cotton, for more carriages, coaches,
and freight wagons, made it necessary to find new sources for
iron, copper, pumice, silica, and mica. There was dire need,
then, for a new kind of professional—a natural philosopher
thoroughly familiar with all the properties of useful rocks and
plants.

Administrative demands were so excessive at Yale University after 1783 that Ezra Stiles never found the opportunity to
add natural philosophy to the courses in electricity, chemistry,
and other sciences he sponsored during his presidency. In
1795, the Reverend Timothy Dwight, a grandson of the great
theologian and revivalist Jonathan Edwards, succeeded Stiles
as Yale's president.

A few weeks before the skeleton of *Mammut americanus*
went on display at the Peale Museum, Benjamin Silliman
asked for a counseling session with Dr. Dwight. The Sillimans
and Dwights had been family friends for decades. Now twenty-
two-year-old Benjamin, a law student, wanted Dr. Dwight's
help in securing a teaching position "somewhere in the South"
so that he might establish a solid reputation and so that he
might save enough money to open a law office.

Dr. Dwight refused. "I have other plans for you," he explained. "If they materialize, you will be of greater service to
our country. America is rich in unexplored treasures. By aiding
in their development, you will perform an important public
service. I wish to recommend you to the trustees as Yale's first
professor of natural history and chemistry." An obedient son of
his faith and clan, Benjamin accepted Dwight's offer.

Hopeful that no son of the Sillimans of Greenfield Hill would embrace the heretical theories of the Philadelphians and Scots, the conservative Yale trustees approved Silliman's appointment, provided his convictions were still acceptable to them after he had completed a year of study at the universities of Pennsylvania and Edinburgh.

"Grace was never said at table nor did I ever hear a prayer in the house," Silliman recalled of the boardinghouse three blocks from the Peale Museum where he took a room early in the fall of 1802. "Our hostess furnished her table luxuriously. . . . Every gentleman furnished himself with a decanter of wine. I do not remember any water drinker at our table or in the house." One of the other boarders was Robert Hare, inventor of the oxyhydrogen blowpipe and soon to become one of the nation's most noted chemistry teachers.

Caspar Wistar soon proved himself Silliman's most eloquent and inventive instructor. Through him, Silliman met Joseph Priestley, the British chemist-inventor whose sympathy with the French Revolution and opposition to Church of England doctrines forced him to flee to Pennsylvania in 1794. "In person he was small and slender," Silliman recalled. "His dress was clerical and perfectly plain. His manners were mild, modest and conciliating so that although in controversy a sturdy combatant, he always won kind regard and favor in his personal intercourse."

Both Priestley and Wistar were supporters of the theory of uniformitarianism, which the Scottish geologist James Hutton had outlined in his *Theory of the Earth*, finally published in 1795. The Peale skeletons of *Mammut americanus* and the other giant bones unearthed in Virginia, New Jersey, and Ohio, Priestley and Wistar contended, not only supported the Hutton theory but also furnished vivid evidence that Earth was far more ancient than 5,806 years.

But Priestley was a Universalist and Wistar an analytical physician, so Silliman left Philadelphia for further study in Scotland with his faith in Congregational doctrine unshaken. He was auditing lectures and interviewing chemists and natural philosophers at the University of Edinburgh when, in a lecture there, the Reverend Thomas Chalmers boldly announced his acceptance of uniformitarianism and challenged the theory of recent Creation.

"There is a prejudice against the speculations of [Hutton and his associates] which I am anxious to remove," Chalmers declared. "It has been said that they nurture infidel propensities. It has been alleged that these philosophers, by referring the origin of the globe to a higher antiquity than is assigned it by the writings of Moses, undermine our faith in the inspiration of the Bible, and in all animating prospects of the immortality which it unfolds.

"This is a false alarm. The writings of Moses do not fix the antiquity of the globe!"

Chalmers's defense of uniformitarianism and its implications of an ancient Earth was momentous, for it was the first frank avowal of its kind by a British clergyman. His declaration set off a chain reaction that spread from pulpit to pulpit throughout the century and would, between 1820 and 1860, influence—perhaps be the primary cause of—the creation of several new religious sects in the United States.

The first responses to Chalmers's "heresy" came from lecturers at Oxford and Cambridge. That winter stentorian defenses of the recent-Creation doctrine echoed at Westminster, York, Gloucester, and Taunton, with the Roman Catholics for once in agreement with the Church of England. The shrillest denunciations came from Methodist chapels; after all, only a generation before John Wesley had thundered: "Before the sin of Adam, there were no agitations within the bowels of the earth, no violent convulsions, no concussions of the earth, no earthquakes, but all was unmoved as the pillars of Heaven. There were then no such things as eruptions of fire; no volcanoes or burning mountains."

The sermons and debates aroused the curiosity of merchants, farmers, and craftsmen; they began to wonder about rocks and what they might mean. Three theories about the Earth's dawn became popular. Each one had a scholar champion.

The theory of *neptunism* was introduced by Abraham Gott-
lob Werner, professor of mineralogy at the mining academy in
Freiberg, Saxony. Werner, who, like Christian Michaelis, grew
up in the ancient mining region of the Erzgebirge in Saxony,
had been a mineralogy major at the University of Leipzig and
had begun teaching there in 1775. It was his belief that the
oceans had been the basic force in forming and shaping all of
the Earth's rock. Rock formed, he said, as mineral precipitates
from ocean water, with some help from volcanic action. As a
native of a region that had suffered through centuries of struggles between Catholics and Protestants, he shrewdly accom-
modated his theory to conform with the belief in recent Creation. Thus neptunism was acceptable even to orthodox Catholics
and Protestants.

Vulcanism, or *plutonism,* also was in harmony with religious
teachings. The folk tales about volcanic disasters throughout the
Mediterranean basin, the enthusiasm for archaeological digs at
Pompeii, the multitude of obvious volcanic cones in France and
Italy all inspired the conclusion that volcanic eruptions caused
by an "upsurge of molten rock from reservoirs deep in the
Earth" were the primary source of the various varieties of rock
and their layering. Christian Leopold Freiherr von Buch, one of
Professor Werner's favorite and most brilliant pupils, was the
major proponent of vulcanism.

Catastrophism incorporated some of the principles of both
neptunism and vulcanism and, since it was the one most easily
reconciled dogma with belief in recent Creation, became the
favorite theory of the clergy. According to the catastrophists,
the Earth had undergone a series of "Judgment Day" cata-
clysms. Each one destroyed living matter and entombed some
of the carcasses in the molten rock precipitated during the
holocaust. The most recent catastrophe had been Noah's Flood,
which, as Genesis recorded, had a few human survivors as well
as a sufficient number of paired creatures to resupply the earth
with animal and bird life.

Catastrophism's outstanding champion was the great French
naturalist, Baron Georges Cuvier. Ironically, Cuvier's extensive

and extremely detailed research in fossils laid much of the foundation for the eventual acceptance of Hutton's uniformitarianism and the subsequent burgeoning of the Age of Science.

Georges Léopold Cuvier, educated at the academy of Stuttgart, began teaching at Jardin des Plantes in Paris in 1795. In the nearby hills of Montmartre, suppliers for glassmakers and soapmakers quarried potash. As they dug deeper and deeper beneath the ruins of ancient buildings, they began to unearth fossil bones. Cuvier contracted for all the bones they found. To make sense of their variety, he developed an orderly system of zoological classification. Coincidentally, he pioneered the technique of reconstructing the softer parts of skeletons via mathematical studies and terrain analyses.

His *Tableau élémentaire de l'histoire naturelle des animaux*, published in 1798, won international acclaim. Two years later, in *Mémoires sur les espèces d'eléphants vivants et fossiles* he announced that twenty-three species of extinct animals had been identified, including the fur-covered elephant that had been the source of fossil ivory in Siberia for centuries.

Rembrandt Peale was touring England, exhibiting one of the *Mammut americanus* skeletons, during the last months of Silliman's study year there. British naturalists recommended Cuvier's writings to Rembrandt. He found them so stimulating that he scrapped the brochure he had prepared in Philadelphia and wrote the far more detailed *Historical Disquisition on the Mammoth*. In a pamphlet advertising his book, he wrote: "In the account of the MAMMOTH, published in October last, the first crude ideas from an imperfect examination were hastily given: Subsequent investigation has discovered many interesting circumstances, which are here detailed; and some passages which were not sufficiently explicit, have been more fully explained. Instead, therefore, of republishing it as a second edition, which has been long called for, it was preferred to give it a more methodical, satisfactory and enlarged form, only adopting such passages as suited the present purpose (instead of the affectation of giving them a new dress), and dwelling upon those parts of the subject which were but slightly noticed in the former publication." A four-page quotation from the "Memorial of Fossil Bones" by "the celebrated Cuvier" was included in the introduction to the *Historical Disquisition*.

Perhaps Rembrandt's use of Cuvier's material initiated the correspondence between the Peales and Cuvier. In any case, the Peales, Dr. Wistar, and Wistar's protégé Benjamin Smith Barton became Cuvier's favorite American correspondents.

In 1804, Cuvier's first year as secretary of the French Academy of Sciences, word reached Paris that the complete carcass of one of the mysterious black-furred *"mammots"* had been discovered in an ice floe on the Lena River in Siberia's far north. An expedition was sent out from Saint Petersburg to excavate it and bring it in for study. Cuvier obtained details about the carcass; then, in 1805, he began assembling data for a definitive paper about the Siberian and American ancestors of the elephant. He received permission to use the Peale drawings of *Mammut americanus* as illustrations. The paper, published in 1806, concluded that *Mammut americanus* was a distinct species, only distantly related to the *"mammot"* of Siberia. It had been vegetarian and had had tremendous tusks that curved out like grappling hooks to facilitate feeding in the swamps and to rake leaves and vines off tree trunks. The teeth of *Mammut americanus* were somewhat rounded and conical. Because of this resemblance to a miniature breast, Cuvier decided that the species should be named *Mastodonte*, or "breast-toothed." The Siberian animals had had huge teeth with shallow ridges running across their flat, broad, crushing surfaces. Because it was giant-toothed, Cuvier reasoned, the beast should retain its Siberian name of *Mammut*, or "mammoth." Cuvier's logic prevailed, although the Peales never conceded. The lumbering creature that had munched the lush vegetation along the banks of the Wallkill in 7000 B.C. became the mastodon; the woolly Arctic giant kept its Tartar nickname of mammoth.

Rembrandt Peale and Benjamin Silliman were still in England when Charles Willson Peale, Dr. Wistar, William Bartram, and others in the Philosophical Society became advisers to the Lewis and Clark expedition. For decades, Thomas Jefferson had dreamed of a naturalists' expedition into the Far West. Now the Louisiana Purchase made it imperative. Thus his orders to Meriwether Lewis and William Clark were far more detailed than simply to "discover a route to the Pacific."

On January 28, 1803, weeks before James Monroe reached Paris as minister extraordinary to France, the president sent letters to Wistar, Benjamin Smith Barton, and Benjamin Rush requesting them to conduct intensive "cram" seminars for Meriwether Lewis at the Peale Museum, the Bartram Gardens, and the vivisection rooms of the College of Medicine.

The springtime schedule set up for Lewis by the president was as demanding as a major military campaign. During March and April, Lewis would commute between Harpers Ferry, Lancaster, and Philadelphia to oversee construction of theodolites, chronometers, and rifles. He would also attend the lectures and demonstrations that would enable him to serve as the expedition's botanist, zoologist, surveyor, geologist, physician, and gun specialist. By May 20, the instruments must be finished, Lewis's lessons learned, and the loaded wagons ready to start west.

The deliberateness of the Harpers Ferry gunsmiths as well as the elaborate precautions to keep the expedition a secret kept Lewis from starting the Philadelphia seminars until early May. He was still there, fortunately, when diplomatic pouches from Monroe and Robert R. Livingston informed the president that the land purchase they were negotiating with Napoleon was not for New Orleans and Spanish Florida but for the largest territory ever negotiated without force of arms: the gigantic Louisiana Territory. This would double, perhaps triple, the land holdings, natural resources, and responsibilities of the United States. Lewis's program of study with the American Philosophical Society was stepped up—the briefings and lectures were expanded to a month of twelve- and fourteen-hour study days.

Benjamin Rush, professor of medicine and clinical practice at the College of Medicine of the University of Pennsylvania, was an authority on Indian social life and health practices. During the assembly of the expedition's chest of medical supplies, Rush interspersed his lectures about purgatives, the proper use of lancets, and the repair of broken limbs with stories about the Indian's sex life, suicide ratio, food preservation techniques, and "the affinity between their religious ceremonies and those of the Jews."

Benjamin Smith Barton and William Bartram took Lewis on lecture tours through the Bartram Gardens and the Peale Museum, then demonstrated techniques for preserving leaves, blossoms, seeds, and tubers.

Caspar Wistar and the Peales became Lewis's tutors in geology, zoology, and the rudiments of taxidermy; the president had instructed Lewis to conduct a search for "the remains or accounts of any [animals] which may be rare or extinct." (Jefferson's use of the word "extinct" again suggests that his repeated public statements denying the possibility of "extinct species" may have been political expediency, rather than personal conviction. Or was he too being converted by the evidence of an ancient Earth being assembled by the uniformitarianists and Cuvier?)

The demonstrations of taxidermy procedures at the Peale studio and the conferences with geographers and mineralogy specialists at the American Philosophical Society were so demanding, and now so important, that Lewis did not leave Philadelphia until June 15. Subsequent conferences with the president and War Department officials in Washington delayed the departure for Pittsburgh until July 4, thus assuring that the expedition could not possibly begin the ascent of the Missouri until the spring floods of 1804 began to ebb.

With this new, unavoidable delay, the president decided to make use of some of Lewis and Clark's unexpected leisure time by sending them to dig for fossils at Big Bone Lick. In the last week of September, Lewis learned that a Cincinnati physician, Dr. William Goforth, was financing a thorough dig at the Lick and was using the water-wheel technique that had been developed by the Peales at the Wallkill dig. Dr. Goforth was certain that medical schools and European museums would be quick to buy up all the fossils he was hoping to unearth.

By the time Lewis and Clark reached Big Bone Lick, Goforth already had an impressive collection of teeth, tusks, and bones. When Lewis explained his mission and the president's eagerness to add Big Bone Lick specimens to the roomful of fossils he had accumulated at Monticello, the physician selected some of the choicest finds and had them washed, waxed, and carefully packed for the three-month journey to Washington. Accounts vary as to the fate of the shipment. One version alleges that the packets were lost en route. Another version claims that the fossils reached Washington, where they were inspected by one of Jefferson's aides and then sent off to a bone yard to be ground into meal. Whatever their fate, the president never saw them.

Soon after Benjamin Silliman returned to New Haven, with assurances to the trustees of his loyalty to Congregational tenets, Lieutenant Zebulon Pike reported to the American Philosophical Society for botany, zoology, and geology seminars. Pike was to command an expedition up the Mississippi in an effort to discover its headwaters and then ascertain the boundaries between the United States and Canada. The War Department considered Pike's assignment almost as critical as the transcontinental trek being undertaken by Lewis and Clark. The massive patchwork of lakes, forests, and prairie between Lakes Michigan and Superior and the Rockies was an unmapped mystery. A few American trappers and traders had ventured as far as the Falls of Saint Anthony and had found the region dominated by Scottish, French, and Indian fur trappers working for the Hudson's Bay Company. In addition to searching for the source of the Mississippi, Pike and his twenty-man party would do surveys, make maps, and observe the regional botany, zoology, mineral outcroppings, and topographic characteristics.

Pike, a native of Trenton, New Jersey, a ten-hour trip up the Delaware from Philadelphia, was a devotee of the Peale Museum and the Bartram Gardens, so he was not as unfamiliar with the subjects to be covered as Lewis had been. Moreover, Rush, Wistar, Bartram, and the Peales had reorganized and streamlined the curriculum they had developed for Lewis. As a result, Pike's briefings took only three weeks.

The Louisiana Purchase and the use of both the Lewis and Clark and Pike expeditions to pioneer field research in natural history provided Benjamin Silliman with a lecture series for the entire first year of natural history at Yale. The blockade patrols by both the British and Napoleon had crippled New England's sea trade, and investment capital was being forced to turn to home industries and development of trade with the West. In addition, once the expeditions were completed there would obviously be intensive migration to the Mississippi valley and beyond, particularly if the disputes with Canada were settled and reasonably safe wagon and ship routes could be established across New York and up the Great Lakes. Both of these factors deepened public interest in the natural sciences. Scores of Yale students began to think about careers in botany, zoology, geology, and chemistry. Thus, Silliman learned, he could excite audiences about "God's wonders" in the intricacies of rocks, soil, water and wildlife, yet stay within the prescribed boundaries of religious dogma. But, like Georges Cuvier, he was honest and thorough in his lectures, even when he realized that the evidence could be construed to give support to the heretical tenets of uniformitarianism. Enthusiastic reports by the ministers and trustees who audited his first lectures brought invitations to address agricultural societies and reading circles in Boston, the Connecticut valley cities, and even remote Maine.

The botanical and zoological specimens brought back from the Missouri, Snake, and Columbia valleys in 1806 kept the Philadelphia specialists absorbed for several years and provided Silliman with more exciting lectures. In December, Meriwether Lewis came to Philadelphia to address the Philosophical Society and answer the many questions that the Bartrams, Rush, Wistar, and the Peales had prepared for him. Neither mammoth nor mastodon nor any beast remotely resembling either one had been sighted during the four-thousand-mile trek, Lewis said. Nor had any Indian related a folk tale that mentioned, even vaguely, a flop-eared giant with tusks.

The only fossil the expedition unloaded at Saint Louis was the "jawbone of a fish or some other animal" that had been found in a cavern near what is now Council Bluffs, Iowa. According to Lewis, William Clark had seen two more giant remnants, but they were too firmly embedded in cliffs, and he was unable to dislodge them. One was the backbone of a fishlike beast forty-five feet long. The other, sighted on the Yellowstone River, was "a rib about three inches in circumference."

Immediately after completing his first expedition, Zebulon Pike was assigned to explore the southern sector of Louisiana Territory, so he did not have time to come east for discussions about what he had observed along the upper Mississippi. But his written reports about the forests in the Sioux homeland, the towering palisades, and the pearl-fisheries served to swell the interest in "God's wonders." Natural history became a fashionable hobby in Atlantic seaboard cities. Silliman's lectures often drew audiences of three and four thousand. Field trips to collect botanical specimens, fossils, rocks, and shells led to the setting up of thousands of parlor display-cabinets and to lively evening discussions and swap sessions in town halls and schools.

For a number of people during this period, the natural sciences became something more than simply an absorbing hobby. John James Audubon began his extraordinary career as a naturalist in Philadelphia between 1803 and 1808. The wealthy, eccentric William Maclure began his lonely explorations of eastern seaboard and Appalachian geology during the same years and in 1809 published *Observations on the Geology of the United States of America;* it included the first quasi-scientific geological map of the area between the Atlantic and the Mississippi. Dr. Samuel L. Mitchill, having decided that natural history was a more stimulating career than medicine, became the first professor of natural history at Columbia College and by 1810 was soliciting support for a Lyceum of Natural History in New York City.

With the changes wrought by international politics, this widespread enthusiasm became economically expedient. The War of 1812 forced the United States, particularly New England, to focus on industrial development. The New Englanders built cotton mills in their river valleys, powered them with water wheels and began to turn the South's embargoed cotton crops into dimity, calico, sailcloth, denim, and sheeting; by 1815 more than 500,000 spindles chattered between New Haven and the White Mountains. On New Haven's outskirts, Eli Whitney began a technological revolution by inventing the tools that made it possible to manufacture rifles and muskets from interchangeable machined parts. This assembly system was rapidly adapted to the mass production of wagon parts, farm tools, the new anthracite stoves, clocks, and even rocking chairs and clothespins.

Now the dream of adding natural-history scouts to that march to the West became necessary for survival. The "heresies" of uniformitarianism were beginning to seem less profane. But there were still two great challenges. First, somewhere, somehow, the Lord must have provided a clue that would explain the topsy-turvy jumble of the terrain He had fashioned. Second, someone had to find a quicker, easier route between New England and the West than that trek to the Ohio headwaters at Pittsburgh. The conquest of one, it would turn out, would provide the key to the conquest of the other.

Soon the Royal Society would acknowledge that an English mining engineer named William Smith had discovered the fossil clues to the Earth's interlaced rock strata. Soon Benjamin Silliman would become coach and mentor to the explorer-teacher who was to be hailed as the "pre-eminent simplifier" of New England's waterway shortcut to the West—as soon, that is, as Amos Eaton got out of jail.

7 | The Simplifiers

In these days Geology became the rage. It was talked on every steamboat and canal packet, and at every public watering place.—Arthur Perry Latham, 1835

BY 1630, the Dutch in Nieuw Amsterdam and the French in Quebec realized that the western wilderness of their colonies was gouged by two long troughs, each of which pierced the seaboard mountains. The northern trough carried the runoff water from the Great Lakes. The southern trough swung gently southeast for almost 200 miles, then received the mountain runoff of the Mohawk River and, another 125 miles due east, jetted it over a cliff into the ocean estuary named for Henry Hudson.

The western apex of the oblique angle formed by the two troughs was a massive waterfall called Oughniagarah ("the Thunderer") by the Indians. It barricaded boat traffic from the four upper Great Lakes.

Pitched battles, massacres, cannibalism, slavery, acute alcoholism, racial hatreds, and political knavery kept this "New York West" a deadly wilderness for almost two centuries. Even after 1783, when the British ceded the mountains and rolling meadow south of the Saint Lawrence to the United States, enmities and political feuds held American migrations "up the Lakes" to a few hundred families. The northern campaigns of the War of 1812, especially the American victory at the Battle of Lake Erie, assured the United States' free access to the Great Lakes, which in turn offered the Northeast a quicker, easier throughway to the Midwestern prairies, Michigan's mineral lodes, and the upper Mississippi valley.

The dream of a ship canal through the Mohawk valley gouge between the Hudson and Lake Erie was almost a century old when, in 1817, Governor De Witt Clinton of New York got the state legislature to approve appropriations for construction of a "grand canal."

This narrow waterway, the Erie Canal, and the railroads that soon paralleled it exerted so strong an influence on the nation's economic and social life throughout the nineteenth century that its role as a source of knowledge for the earth sciences was ignored by people recording the "rise of American civilization." Yet, just as New York City quickly succeeded Philadelphia and Baltimore as our largest seaport and New England became a prime supplier for and investor in the West because of the canal, the area between Albany and Buffalo soon rivaled Philadelphia as a center for natural-science research and education because of the rock strata, soils, and flora along the canal route.

Because so many of the advances that took place over the next several years were attributable to one extraordinary genius, geologists still refer to the decade between 1820 and 1830 as the Eatonian Era. The educational foundations created by Amos Eaton influenced the careers of youngsters who, in the latter half of the nineteenth century, became veritable giants in the fields of geology, paleontology, and other natural and physical sciences. The canal route, with its rich and varied supply of fossils, rock strata, soil types, and topographical features, became Eaton's laboratory for scientific research and a training ground for the people who would be instrumental in establishing the national parks, organizing the first geological surveys and the first dinosaur digs in the Far West, and opening up the petroleum industry.

Amos Eaton was born May 17, 1776 in a slab-and-stone farmhouse in the western foothills of the Berkshire Mountains. The area was creekside pasture and stump-pocked cropland a day's ride by oxcart from the century-old village of Kinderhook, on the Hudson. Eaton's father, Abel, was a first-generation New Yorker who had emigrated from Connecticut during the French and Indian Wars. In 1775, Abel married seventeen-year-old Azubah Hurd and began improving the eighty-acre farmstead she had inherited. About the time of Amos's birth, the march by General Washington and his Continentals from Boston to New York City lured Abel. He was appointed a sergeant of foragers and proved so competent that he was promoted to captain and followed the new Stars and Stripes from Harlem Heights to Yorktown.

Abel sired two more sons during the war, and, since both were born in wintertime, it can be assumed that his furloughs were co-ordinated with springtime plowing and seeding. But the grubby tasks of "making the crops" and caring for the livestock were left to his young wife and their first-born.

The community, eventually named Chatham, had much in it to stimulate a toddler's curiosity about rocks, waterways, and forests. The tides of Paleozoic time and the geologically recent glaciers had created foothills of grit against the limestone, slate, and granite backbone of the Berkshire and Taconic mountains. The topsoil contained enough limestone to encourage the growth of forests of chestnut, oak, and white pine on the slopes and hundreds of annual and perennial plants in the bottom lands. The intricate ecological relationships young Amos observed—from the bee's and wasp's roles in pollination to the nut tree's dependence on the frugality of squirrels and chipmunks—brought more questions to the boy's mind than his mother and grandmother could answer.

There was a Presbyterian church in Chatham. Its pastor was a Yale graduate who owned a roomful of books. Amos became such a devotee of the pastor's library that, in 1790, he was chosen to deliver an oration at the village's Independence Day celebration. A few days later, his knapsack on his back, the fourteen-year-old trudged off toward Duanesburgh, near Schenectady, to be apprenticed to a cousin who was a blacksmith and surveyor.

The apprenticeship lasted sixteen months and provided Amos with skills that would finance the college studies he was now determined to undertake. Most of the money he earned as a surveyor and journeyman smith in and around Chatham during the next three years went to his minister in exchange for tutoring in Latin, Greek, geography, logic, natural philosophy, and

mathematics. At nineteen, the minister and other village dignitaries felt that he was worthy of hire as teacher of the village
school.

This was one of the briefest chapters in his career. After one
term as a teacher, he repacked the knapsack and crossed the
Berkshires to begin studies toward the A.B. degree at Williams
College. He took his deerskin bundle of surveying instruments
with him; over the next four years he earned enough at surveying to cover his board and his tuition fees.

The mysteries of natural philosophy continued to fascinate
him. The profession of land agent would make use of his surveying skills and offered the opportunity to do research in the
field that interested him so much. But to be a land agent, he had
to have a knowledge of law as well. Soon after graduation from
Williams, in 1799, he became a student-clerk in the New York
City office of Josiah O. Hoffman, the state's attorney general.
He also enrolled for courses in botany at Columbia College.

Eaton's eagerness to learn and his persistent efforts to re-
phrase legal as well as natural-philosophical jargon in simple
terms "that can be understood by all who read" intrigued
another student-clerk on Judge Hoffman's staff—seventeen-
year-old Washington Irving. Irving was considered "the boy
wonder" of the firm, but, he confided to Eaton, he was studying
law partly because his older brothers insisted on professional
training, and mostly because he was in love with Judge Hoffman's daughter, Matilda. "The law," Irving told Eaton, "wor-
ries me prodigiously, in spite of all my resolution . . . my
great ambition is to write . . . any form of writing that will
enable me to create on paper."

After three years of study and apprenticeship in Hoffman's
office, Eaton, on October 29, 1802, passed the state bar examinations. He and Washington Irving had won the friendship of
William Van Ness, a successful attorney. During the 1790's,
Van Ness built a mansion home at Kinderhook and became the
legal adviser to the Livingstons, Van Rensselaers, and other
wealthy and influential mid-Hudson valley families. Van Ness

persuaded William Livingston, a miserly cousin of the states-
men Robert R. and Edward Livingston and the owner of farm
and forest tracts in the Catskill mountains, to hire Eaton as land
agent for these properties.

Eaton moved to the village of Catskill during the spring of
1803 and began the onerous task of pressuring tenant farmers,
lumber contractors, leather tanners, millers, and quarrymen for
the rental fees that, Livingston seemed endlessly to whimper,
"will save me from bankruptcy." But the "nag-and-threaten"
trips up the mountains also offered opportunities to explore the
area's geology and botany.

By 1808 Eaton's home was a miniature museum of labeled
and indexed rock specimens, dried flowers, herbs, and marine
fossils. The surrounding gardens were an arboretum of flowers,
herbs, weeds, and shrubs native to the Catskills and to the
Delaware and Susquehanna headwaters. The thousands of
shells and other fossil marine animals freckling the black cliffs
of the Delaware headwaters near Middleburgh were so exciting
that the Mohawk Dutch and Irish tenants there began to accept
"that snooper of Livingston's" as a harmless and "pretty interesting" fellow; they began saving good fossil finds for him.

Vrooman's Nose, as some Mohawk Dutch with a sense of
humor had named it, and the other cliffs palisading the table-
smooth valley at Middleburgh were the most puzzling parts of
all in the Catskills' geologic jumble. It was heresy, of course, to
wonder why God, that week in 4004 B.C., had piled this brick-
red sandstone on flaky chocolate shale on Quaker-gray limestone, then twisted them into taffy layers and studded each
layer with distinctive forms and types of shells, fish, and
crustaceans. The same sort of cake layering appeared in creek
gorges on the eastern slopes of the Catskills. Was this coincidence? Could it be evidence of some occurrence that somehow
had been omitted from the translations and retranslations of the
Bible? That Wallkill bog where the Peales recently dug out the
mastodon was only a hundred miles away, at the southern base
of the Catskills. Was there a connection, somewhere in the

Earth's past, that linked the elephants in the Wallkill, the marine fossils in the Schoharie, and those giant leaves and stems they kept digging out of Pennsylvania's anthracite?

The yellow fever epidemic that swept up the Atlantic seaboard during the summer of 1809 caused Eaton and Washington Irving to become trans-Hudson neighbors. Irving achieved his first success as an author in 1808 and Matilda Hoffman had consented to marry him. But "the fever" killed her a few weeks before the wedding. William Van Ness urged the grieving author to leave memory-laden Manhattan and come to Kinderhook as tutor for his children. Eaton and Irving visited one another frequently that fall and winter. The detailed topographical descriptions in Irving's *Rip Van Winkle* may be taken as evidence that Irving accompanied Eaton on at least one rent-collection trip up the Catskill "cloves," just as Irving's natural-history discussions and wanderings with the Kinderhook schoolteacher Jesse Merwin are immortalized in his characterization of Ichabod Crane in *The Legend of Sleepy Hollow*.

Irving was one of the first to learn of Eaton's intention to resign as Livingston's agent and become an independent realty attorney. And they discussed Eaton's dream of undertaking a radical experiment in education by starting a botanical institution in the village of Catskill. In 1806, with loans from his father and a brother, Eaton had made the initial payments on a thousand-acre tract of Catskill farmland owned by Nathaniel Pendleton. Pendleton was the arrogant New York City attorney who had served as Alexander Hamilton's second in the fatal duel with Aaron Burr. There were ugly rumors that he had fleeced Catherine Greene Miller, widow of both General Nathaniel Greene and Eli Whitney's partner in the cotton-gin fiasco, out of much of her inheritance. But the terms he offered Eaton seemed lenient, and he professed deep interest in Eaton's geological and botanical research.

Assured that the thousand acres could be used as negotiable security on other investments, Eaton left the Livingstons and opened an office. His economic horizon was so promising by the spring of 1810 that, on April 19, he placed the following advertisement in the county's newspapers:

"The subscriber, having been frequently solicited to teach (mathematics and botany) in this village, has made such arrangements in his business as to be able to attend to A Botanical Institution every Monday and Saturday through the months of June, July and August and perhaps longer if sufficiently encouraged. The plan of arrangement proposed is to attend to surveying, navigation, gageing, gunnery, Algebra, astronomic calculations and all other branches of mathematics five hours in the forenoon; and Botany four hours in the afternoon. All persons who understand the rule of three will be attended to in the mathematics. And all persons of both sexes, from twelve years old to sixty, if required, will be instructed in the art of Botany. Although there is no botanical garden in this neighborhood, plants and flowers will be secured, sufficient to illustrate every principle."

So many people applied for the botany course that Eaton wrote his former botany professor at Columbia, Dr. David Hosack, and asked his advice about lecture sequence and displays. Dr. Hosack promptly offered his services, free of charge, as consultant to the institution and congratulated Eaton on being "the first in the field." He prophesied, correctly, that "your example will be followed by many, if not most, of the academies throughout the state. This will result in the discovery of many valuable plants which at this moment are trodden underfoot as unworthy of regard."

The demonstration lectures on Catskill flora and their care and uses were so interesting that by the end of the first month three attorneys, three judges, the Presbyterian minister, the postmaster, and a printer had formed a committee with the primary purpose of "engaging Mr. Eaton at our own expense to teach all persons of any age, of a decent character and suitable education, for two months gratis." On July 17, the committee paid for a "Publick Notice" in *The Catskill Recorder* announcing that they "have it in contemplation to build a commodious greenhouse next spring, and to make a suitable collection at their own private expense."

Eaton became so stimulated by prospects for the Botanical Institution that he decided to sell fifty acres of the land he was purchasing from Nathaniel Pendleton and use the profits to buy equipment. Pendleton refused to co-operate and, without warning, filed charges that Eaton had forged one of the documents of the agreement made in 1806.

The Hudson valley patroons had secured their economic position over the past 150 years by skillful lobbying and shrewdly drawn up marriage contracts. As a result of their influence, the New York penal code was harsh in its protection of landholders' rights. Conviction on a charge of forgery that involved a property sale called for a sentence of "life imprisonment without clemency." The only way of avoiding the sentence was to win a pardon or an order of banishment from the governor.

Despite the testimony of friends in Catskill, character statements from Columbia professors and William Van Ness, and Pendleton's admission that he had agreed to the sale in 1806, Eaton was found guilty on August 26, 1811, sentenced to life imprisonment, and shipped in chains down river to New York City's Newgate Prison.

Newgate loomed over four acres at the Hudson River end of Tenth Street in Greenwich Village. Its granite enclosing wall was fourteen feet high on the village side and twenty-three feet high on the riverbank. The prisoners' daily rations, budgeted at a wholesale value of five cents per prisoner per day, included oxheads and hearts, lamb plucks, salt pork, black bread, molasses, and infrequent beans. The one-story brick workshops barricading the exercise yard were production centers for hand-loomed blankets and cloth, thread and yarn spinning, shoe-making, brush and broom finishing, blacksmithing, and carpentry.

Workdays began at 6:00 A.M., slowed for a 30-minute lunch, then resumed until 6:30 P.M., when guards inspected the shops and marched the prisoners off to the mess hall. The march back to the cells began at 7:30 P.M. Convicts on "good behavior" lodged three to a cell. No lights were permitted. The only heating unit in each cell block was a Franklin stove alongside the guard's desk and chair at the doorway. Prisoners who dared to complain, to attempt escape, or to doze or act "inattentive" during one of the tedious Sunday sermons went into the block of filthy solitary wells, were denied exercise and conversation, and were given only stale bread and water.

After Governor Daniel D. Tompkins denied the pleas of both Judge Hoffman and William Van Ness to pardon Eaton, they shifted their attention to William Torrey, Newgate's warden. Judge Hoffman was recorder of New York City's Common Council; Torrey was one of the nine councilmen. Hoffman knew that Torrey's teen-age son, John, was a zealous student of botany and geology and had shelves of specimens in his bedroom. During a Council meeting, Judge Hoffman casually asked Torrey about his son's latest finds, then described Eaton's background and his catastrophic experience with the Catskill Botanical Institution.

"I'll keep an eye on the fellow," Torrey promised.

A week or two later a guard told Eaton to fall out of line and ordered him to the administrative office of the workshops for "tryout" as an accountant. His work was inspected each afternoon by a deputy warden; his tally sheets were questioned and readded. These inspections continued for a month. He was told that in the future he could dine with the guards. After two months as the accountant, an order came to report to the warden's office.

Warden Torrey had pushed his chair back from his desk. His legs were crossed and he was smoking a long cheroot. An index finger drummed against a chair arm as he studied Eaton's face. Behind him stood a slender youth with deep-blue eyes, hair that curled fashionably to the shoulders of his fawn-colored coat, and long, slender hands.

"My son, John," Torrey grumped after a moment. "He wishes to discuss the Linnaean method of analysis with you."

For the next hour and a half, Torrey sat wordlessly puffing cigars while the youth cross-examined Eaton, first about the late Carolus Linnaeus's system of bionomical classification of plants and animals, then about the plant life and rock strata Eaton had researched in the Catskills.

When the bell sounded for dinner, the warden flicked his third cigar butt into a spittoon, sighed, and growled, "I suppose my back office will be the best place to continue this nonsense. I shall advise you of the boy's wishes, Eaton."

"Now I am permitted to remain in the Agent's back office until bedtime," Eaton wrote his wife that winter. "With this privilege are connected many others, which savor of liberty. I not only walk out in the yard when weary of sitting, but can occasionally go without the gates with the keepers along the bank of the river &c on Saturday afternoons. I spend my evenings in progressing with my Botanical work, which I commenced in Catskill. I have contrived a new method of arrangement, by which I can exhibit all the known species of plants (about forty thousand) in one small duodecimo volume. So that you can readily determine the name of every plant and its uses in medicine, diet, agriculture and the arts by merely inspecting the flower and some few other parts. The Agent's son, John Torrey, supplies me with all the books I want, which are not in the State-prison library."

The two-year working relationship with John Torrey flowered into Amos Eaton's most abiding friendship. Its scope would be summarized by Eaton in his journal entry for August 11, 1838, when Torrey was professor of botany at Princeton University: "On this day," he confided, "I received the first numbers of the first volume of a work of most important expectations. It is a first attempt at a full Flora of North America; and arranged according to the Improved Natural Method. The author is Dr. John Torrey, M.D. This number was published . . . a quarter century after the very month in which I taught the author the name of Calyx, Corol, Stamen and Pistil in the Artificial Classification of Linnaeus, and the Linnaean method of Analysis. I was then in a state of deepest affliction in which perjury and bribery, stimulated by avarice, could subject unconquerable integrity. Dr. Torrey was then a young scion of great promise, whose ruling passions were generous sympathy and zeal for

Natural Science 'for its own sake.' By him, were my afflictions greatly alleviated; and to his operations on the mind of his venerable father's sympathies, I am, in part, indebted for relief from my powerful oppressors."

Perhaps through the intercession of Professor Hosack, but more probably because of William Torrey's proud growls about "that convict tutor of my son's," Professor Samuel Mitchill of Columbia also became a habitué of Newgate's "back office." He was winning his long appeal to the Columbia trustees for the creation of "a museum like the Peales" and was at work on the first American translation of Cuvier's *Theory of the Earth*. In 1813, he became a consultant for the book Eaton would eventually publish under the title of *A System of Botany for the Northern States* and a second volume, 342 pages long, called *A System of Mineralogy Being the Essential Part of Kirwan's Elements of Mineralogy, Corrections and Improvements from Accum's Chemistry*.

The stacks of manuscript that began issuing from William Torrey's back office impressed Torrey almost as much as the fact that the Newgate account books had never been so neat or concise. He bragged about Eaton so often at City Council meetings that Mayor De Witt Clinton asked Torrey to bring his prize prisoner over to City Hall for a chat. The initial meeting took place during the winter of 1813–14. Eaton's combination of lucidity and scholarship about natural history was exactly what the mayor sought for the postwar gubernatorial campaign he was planning. There were many competent scholars available at Columbia, Yale, and Princeton, but very few of them were able or willing to impart their learning to the farmers, butchers, millers, and carpenters—and theirs were the votes that decided elections.

In 1810 Clinton had been a member of the state commission sent up the Mohawk valley and on to that gloomy gorge of Niagara Falls to assess the feasibility of a grand canal between the Hudson and Lake Erie. The canal could be built easily enough, but, Clinton learned during the weeks on the Niagara frontier, Henry Clay and the other "war hawks" in Congress had grossly underestimated the military strength and loyalty of the Canadians and their Indian allies. An American invasion of Canada, he foresaw, would not be the triumphant march that the war hawks predicted.

The disastrous defeats at Detroit and on the Niagara validated his fears. Not until Rhode Islander Oliver Hazard Perry and some select roughnecks from the Brooklyn Navy Yard won the Battle of Lake Erie was there even assurance that the United States would be able to retain any rights to the Great Lakes. Now that William Henry Harrison's Kentuckians had recaptured Detroit and seriously crippled the Indian confederacy at the Battle of the Thames, it seemed probable that this war would end with the New York–Canada boundaries about where they were in 1812. It was possible that those Southern generals might lose everything with their haphazard tactics in the Saint Lawrence valley. If they didn't, a sort of "freedom of the Lakes" was in prospect. If this happened, then construction of the grand canal would be feasible.

The battle to win appropriations for construction of the canal, Clinton realized, would be difficult and unpleasant. In order to win, it was essential to gather information about the minerals, soils, farming potentials, and industrial opportunities along the three-hundred-mile strip the canal would traverse. Amos Eaton's obvious genius for exciting laymen about such ponderous subjects as geology, botany, and zoology might prove to be a valuable asset. Eaton became an occasional visitor at the mayor's office and home as the war moved toward the Niagara–Saint Lawrence stalemate.

The War of 1812 ended on Christmas Eve, 1814. But frenzied construction of warships along both shores of Lakes Ontario and Erie underscored the urgency of a naval-limitations treaty to assure freedom of the Lakes. While Washington and London moved ponderously toward the Rush-Bagot Convention that ended the arms race, De Witt Clinton marshaled the legislative forces that, during the fall of 1815, would introduce a bill for initial canal appropriations. It was obvious that Clinton would build his gubernatorial campaign around the issue of the canal. Governor Tompkins, wearied by four years of war and the eternal bickerings of the state militia, yearned to return to the dignity of a federal office. Possibly to avoid charges of bias in the 1816 election campaign, possibly to placate Clinton, Torrey, Van Ness, and Hoffman, Tompkins issued an order on November 17, 1815, that "within three months from the date hereof" Amos Eaton was to depart from the state of New York and "never thereafter return to same."

Either the Torreys or Clinton told Eaton in October that the order was on the governor's agenda. All of the negotiations for Eaton's enrollment in Benjamin Silliman's natural-history courses at Yale were completed before December 1. It seems probable that the word came from Clinton, and that some arrangement was made to advance enough money to finance the Yale studies; Eaton's legal fees during those last weeks at Catskill had wiped out all of his savings. Mrs. Eaton and the children had been forced to return to the Abel Eaton farm at Chatham, where her parents did not fail to point out "the evil heritage of our grandchildren."

Eaton's decision to keep a daily journal of every mineral outcropping and rock stratum he could identify on his trip home and thence east to New Haven also indicates that he had reached an agreement of sorts with Clinton and was making notes for lectures and books.

William Torrey gave his best bookkeeper freedom on December 2, 1815. Eaton had been a prisoner for four years and four months. He was home for Christmas, continued his investigation of the Berkshires' marble, mica, pyrite, granite, gneiss, sandstone, and hornblende ridges through January blizzards, and moved his family across the state line to Massachusetts before the governor's three months of grace expired on February 18.

At New Haven, Eaton approved the small home with garden that his wife had selected, unpacked his books and rock samples, and began day-to-evening studies and discussions at Yale. Mrs. Eaton and the children settled in, spaded and planted the garden, and pondered proper playmates and a summer schedule of chores and studies. Once these matters were settled and her family's social future once again appeared promising, Mrs. Eaton died of a heart attack while preparing dinner on July 23, 1816. She was thirty-two years old.

A week after the funeral, the widower began a cautious survey of New Haven matrons. He was as thorough and dispassionate as he was in analyzing rock samples, and finally approached a robust, placid widow named Ann Bradley. Mrs. Bradley consented to become the Eatons' housekeeper. After another four months, Eaton was satisfied. He married Mrs. Bradley.

Now Eaton was forty years old, in an era when the average life span was thirty-five. His schedule at Yale was heavier than his work load had been at Newgate. Professor Silliman lectured on chemistry each morning, Monday through Saturday, and on mineralogy on Wednesday and Saturday afternoons. Five afternoons each week, Eaton studied and made notes about the Colonel George Gibbs Collection of minerals, which had recently been presented to Yale and was being temporarily stored in Silliman's office. Most evenings were devoted to editing sessions with Silliman's associate, Professor Eli Ives. Ives had consented to sponsor and edit Eaton's manuscripts about botany. Both *A System of Botany for the Northern States* and *Richard's Botanical Dictionary* were completed during these evening sessions. Both books were dedicated to Professor Ives, but neither credits Eaton as author or editor.

De Witt Clinton won the gubernatorial election of 1816. Eaton would have to be able to offer more than ghostwriting abilities in order to be asked to return to New York when the struggle for the grand canal appropriations was won. In November, soon after the election, Eaton rode a sloop up the Connecticut River to Windsor then changed to one of the Concord coaches that crossed the Berkshires to Williamstown. In his carpetbag were letters from Professor Mitchill of Columbia and Professors Silliman and Ives of Yale attesting to his competence as an instructor in mineralogy, botany, and zoology.

President Z. W. Moore of Williams College had reservations

about having an ex-convict, even if he was a Williams graduate, lecture at the school, but he consented to a trial series of lectures, to be monitored by a faculty committee, on "elementary and practical mineralogy and botany."

The lectures were so succinct, and Eaton's use of household utensils as illustrative tools so original and effective, that he was unanimously approved to initiate the courses for both faculty and upperclassmen in the spring, 1817, term. The courses began in April and met six mornings a week. "I become all things to all men," Eaton boasted in a letter to John Torrey. "I turn everything in science into common talk. I illustrate the most obtuse parts by a dishkettle, a warming pan, a bread tray, a tea-pot, a soap bowl or a cheese press." By mid-term, professors as well as students were lugging kitchenware and garden tools across the campus to the lecture hall to explain how each item could be useful in demonstrating some principle of plant behavior or geology.

A few days before the September commencement, President Moore suggested that Eaton acquire an academic gown for the occasion—the trustees had approved both an honorary master of arts degree for those "common talk" lectures and a permanent faculty appointment as lecturer of the natural sciences.

But the real honor came from Albany. It was dated September 15, 1817. The letter, written by Governor Clinton himself, stated that "Amos Eaton has devoted himself to the instruction of youth and the Cultivation of Science and is a fit object for our mercy. Therefore know ye that we have pardoned and do by these presents pardon him unconditionally and absolutely from the felony aforesaid and of and from all sentences, judgements and executions thereon."

The challenge had come. But first Eaton must test his "common talk" approach on nonacademic audiences. One of the Williams students who had been excited by the botany lectures had already volunteered to solicit subscriptions for a seven-week lecture series at the town hall of his home town, Northampton, Massachusetts. Governor Clinton's pardon influenced Eaton not only to accept the Northampton challenge but also to experiment with another idea in educational communication. He invited all of the Williams students and faculty who could find the time to join him on a fifty-mile hike down and across the Berkshires to Northampton. Each hiker was to bring a rock hammer and a knapsack in which to collect rock samples. The purpose of the hike would be to "make geometrical surveys of the countryside and prepare a geologic map of the area fifteen miles in breadth."

There were thirty-five volunteers. They were so enthusiastic that their meandering route was more than two hundred miles long. Most of the hikers had relatives or friends living along the dirt roads, so the party had free lodging in haymows or barns and, Eaton wrote John Torrey, "I spent but three dollars."

The Williamstown–Northampton survey is the first recorded instance of American use of the field-trip technique that would become such an important tool in university and public-school curriculums a century later.

The "common talk" lectures at Northampton proved as popular as they had been at Williams. "I have given a short course of evening lectures on chemistry here," Eaton reported to John Torrey. "I exhibited all of the usual experiments with gases, acids, salts &c. Then I exhibited minerals enough for understanding geology. . . . My classroom is crowded with the first people here of both sexes and all ages. I have fifty-five ladies, four practising lawyers, three students at law, three practising physicians, two students in physics, three other gentlemen from Yale College, one senator, one representative in Congress, one Common Plea judge &c."

Using "a pewter sucking bottle for my fluoric gas bottles, a stone jug and a tin tube [as my] earthen retort, a teakettle with the cover luted on [as] my iron retort," Eaton introduced "the first people" of Belchertown, Worcester, Monson, Brimfield, and other Massachusetts communities to the wonders and economic potentials of their plants and rocks. "What is botany good for is asked by every parent, guardian, etc.," he wrote Torrey. "If this answer is given that it teaches the *virtues* of plants, the subscription is readily filled. If you speak of its advantages in sharpening the faculty of discrimination, being a pleasing substitute for frivolous or mischievous amusements, etc., etc., you are laughed at . . . and get no students."

He was being equally subtle in his relations with the clergy. The evidence of fossils and rock strata convinced him that the Earth was, in terms of human comprehension, a very ancient planet. But he carefully avoided discussions about "the age of the Earth" and the "awesome influences of Noah's Flood," although he gave special passes to his lectures to clergymen and asked for their opinions at the end of each series. With the instinct of an evangelist, he knew that "conversion" to aware-ness of the impact of botany and geology on everyday life was the soundest way of encouraging a re-examination of the tradi-tional interpretations of Genesis and "the immensity of God's Universe and Will."

Governor Clinton was just as subtle in reserving Eaton as his ace in the hole for the spring, 1818, session of the state legisla-ture. The initial appropriation for the grand canal merely ap-proved a barge channel four feet deep between the Hudson River and Lake Ontario. The massive Niagara barrier to the Upper Lakes remained and the cartmen and toll-road operators at the Niagara portage were of course lobbying to keep it that way. During the 1818 legislative sessions, Clinton believed, enough economic realism could be marshaled to assure a ma-jority for a Lake Oneida–Lake Erie extension of the canal. Eaton's role in this would be to give a series of those "common talk" lectures about the geology and botany of western New York and its potential for agricultural, industrial, and urban development. In February, when Eaton received the invitation to deliver the series, he canceled all his Massachusetts engage-ments, packed his kitchen utensils and other paraphernalia, and took the stage down to New Haven to celebrate the "invitation to come home."

The lectures were delivered at Albany's Society of Arts in April and May. The first three were "free to all comers." Again, Eaton's wizardry with "common talk" and his use of the kitchen utensils to demonstrate scientific principles were effective. The appropriation to extend the canal to Lake Erie via Syracuse, Rochester, Lockport, and Buffalo did squeak through the legislature. Moreover, Eaton common-talked himself into the most momentous relationship of his career.

At the close of one of the lectures, a delegation of physicians, lawyers, and clergymen from Troy introduced themselves. They represented a group promoting the founding of a Lyceum of Natural History in Troy, and they wished to underwrite a minimum guarantee for a summer-long series of lectures. The wealthiest sponsor of the series and of the campaign for the lyceum, Eaton soon learned, was the politically powerful patroon Stephen Van Rensselaer.

The Van Rensselaers had been the grand aristocrats of Eaton's childhood. For fifty miles north and south of the Hudson-Mohawk junction, they retained the "perpetual fief of inheritance" granted to Arendt Van Rensselaer in 1629. Helderberg, Berkshire, and Catskill farmers continued to pay the family ten bushels of wheat per year for each hundred acres of land they worked. The development of Troy, Watervliet, Cohoes, Rensselaer, Hudson, and other communities in the upper valley further expanded the Van Rensselaers' financial empire.

Stephen, eighth patroon of the manor, was a grandson of the mid-Hudson patroon, Philip Livingston, and was married to a daughter of the Revolutionary hero and upper Hudson patroon, General Philip Schuyler. He was a graduate of Harvard College, had been a lieutenant governor of New York, and was a commissioner of the grand canal. Both his prestige and political career had been crippled during the War of 1812 when he accepted the command of the New York militia on the Niagara frontier. His failure to win the co-operation of Federal troops and to discipline the insolent company and regimental commanders of the militia was responsible for the disastrous American defeat at the Battle of Queenston Heights. Van Rensselaer promptly resigned and ever since had stayed away from politics, concentrating instead on agricultural improvements and his realty interests.

He was a corresponding member of Philadelphia's famed Society for the Improvement of Agriculture and a founder of the Albany Agricultural Society, and he advocated the cultivation of such unorthodox crops as Lucerne grass (alfalfa) and improved strains of the sweet corn discovered growing in Onondaga and Seneca fields in 1779 during the brutal Clinton-Sullivan invasion of the Five Nations' homeland.

The response to Eaton's first Troy lecture was so enthusiastic that Van Rensselaer showed up for the second one. He soon persuaded attorney William L. Marcy, a future governor of New York, to subscribe too, and he led the group campaigning for Eaton's appointment as permanent lecturer and chief collecting agent for the new lyceum's museum of minerals, fossils, and upper Hudson botanicals.

Before the lecture series ended, Governor Clinton and Van Rensselaer were pledging financial support for a second Albany lecture series, to be delivered during the winter months, exclusively for members of the legislature. They asked that these talks deal with "the specific applications of geology and chemistry to agriculture." Eaton quickly found a home in Troy, then hurried back to New Haven to pack his beloved specimens, books, and household goods for the wagon haul over the Berkshires. The family preceded the freighters in a buggy. The boys did most of the driving down the west slope of the mountains and were ordered by their father to make long detours. This was Hudson watershed country, and Eaton was using the journey to search out choice specimens for the Troy Lyceum.

The legislature lectures were delivered three evenings a week in the Society of Arts Room at the State Capitol. More than half of the state senators and representatives, as well as scores of bureaucrats and lobbyists, became regulars and took notes about soil tests, crop-field enrichment, and chemical reactions. The information would be useful to their constituents—and in future speeches. "Now," Eaton gloated in a letter to John Torrey, "I shall have collectors of natural substances spread over most of the state. . . . I have learned to act in such a polymorphous character that I am to men of science a curiosity, to ladies a clever schoolmaster, to old women a wizzard, to blackguards and boys a shewmen and to sage legislators, A Very Knowing Man."

A certificate of honor was presented to "the master teacher" at the end of the last lecture. A week later, the legislature passed the act for New York's first appropriation (ten thousand dollars) "for the promotion of agriculture and industry." The appropriation and the subsequent founding of a State Agricultural Society preceded the federal government's creation of a Department of Agriculture by four decades. Thus Amos Eaton's muffin-tin and tea-strainer demonstrations to the New York politicos in 1818 and 1819 in their own way laid the groundwork for the post–Civil War development of land-grant colleges and agricultural extension programs. This was an important step in the transition from natural philosophy to practical science in America.

The legislature lectures probably contained some guarded references to *stratigraphy*. This system of identifying rock strata by their highly individualistic colonies of fossils answered questions that had plagued Eaton since childhood, and it was logical that he would be eager to translate its implications into "common talk." References to stratigraphy began to appear in British journals in 1816. Credit for solution of the rock-strata puzzle initially went to George B. Greenough, wealthy president of the Geological Society of London. Years later, however, Greenough admitted that he had "borrowed" much of his data from the *Geological Map of England and Wales*, published in a four-hundred-copy edition in 1815. The map was the product of twenty years of research by William Smith, a canal engineer at Bath. Like Eaton, William Smith was a self-made scientist, endowed with relentless curiosity about the patterns of nature. He too came from a family of modest means, attended common schools, and became a surveyor's apprentice in his teens. And he possessed the urge to simplify. Eaton simplified with words. William Smith simplified with maps.

As a surveyor, and later as an engineer on the canals being dredged between the coal lodes of Somerset, Smith noted that each kind of rock was characterized by a distinctive array of fossils. From 1796 through 1811, Smith traveled from job to job throughout England and Wales in order to study the consistency of the rock and fossil patterns. By 1811, so many of his associates in the Bath Agricultural Society were convinced that rocks could be identified by their fossil content that a fund was raised to finance Smith's preparation of a colored *Geological Map of England and Wales*. The map eventually brought Smith acclaim as "the father of English geology" because the stratigraphy it introduced became the basis of the geological timetable that geologists and paleontologists drew up between 1820 and 1900.

"The decade 1820–29 may be called the Eatonian Era," George P. Merrill of the U.S. Geological Survey concluded in his definitive *The First One Hundred Years of American Geology* (1924). "It was, so to speak, a transition period; one in which the possibilities of state and governmental surveys were seriously considered and one, too, in which, as far as America was concerned, there was made the first systematic attempt at correlation by means of fossils. . . . The cosmogonists had largely disappeared and in their places were men who had first learned to observe and then to draw conclusions according to their understanding of the observed phenomena."

Then, as now, politicians were so intimidated by the voting power of the pious that any direct reference in the legislature lectures to Smith and his "simplification table" could have proven catastrophic. In the summer of 1819, however, Eaton was once again lecturing in Hudson valley and Berkshire communities, and here he did freely discuss Smith's work and make use of his techniques.

Eaton's eagerness to apply Smith's theory to the fossil wonderland on the west slope of the Catskills brought him back to the Schoharie and Delaware headwaters the following fall. The trilobites, shells, corals, crinoidea, and other Paleozoic creatures he carted home from the Schoharie became demonstration pieces for the semiweekly lectures he delivered at the Troy Lyceum between November and April. During these months, he also was hard at work on a revision of a textbook on geology. Early in April, 1820, he recrossed the Berkshires to deliver lectures on botany at Lenox, Old Stockbridge, and Great Barrington, Massachusetts. Then he rode on to Castleton, Vermont, to give a six-week lecture series at Vermont's first school of medicine, Castleton Medical Academy, where he had recently been appointed professor of botany.

On the way home in mid-July, he thought about the first geological survey of Albany County—if it was to be done at all, it had to be completed before the fall term began at Castleton. He presented his field work, analyses, and maps, along with his bill, to Van Rensselaer in the last week in September; he was on campus at Castleton before October 1. "I had more to do here than I expected," he wrote John Torrey in December. "I have given 109 lectures. I had not only to give my own parts, chemistry, botany and natural philosophy, but I have given a full course of medical jurisprudence and geology, with the elements of mineralogy. Now I have done and give *all* my time to botany. . . . We have forty-four very likely young men. . . . I shall leave Castleton in three weeks. I calculate to stop for one month at each of the following places before I arrive at Troy: Rutland, Manchester, Bennington, Vermont and Pittsfield, Massachusetts."

Overwork, infected well-water, and Franklin-stove heating caught up with him in Rutland, Vermont, at the end of December. The "incipient typhus," as local physicians called it, invalided him for two months. "Here at a friend's house," he wrote on February 21, "I intended to shut up in a closed room and finish the 3d Ed. of the Manual. . . . But a most alarming Cyranche (as doctors call it) attacked me. During thirty-eight days, I have been able to write part of two pages."

The enforced leisure had its advantages—he had a chance to review his progress and redirect his goals. The appeal of his lectures to housewives, merchants, farmers, and preachers clearly indicated that applied botany, geology, and physics could play vital roles in American homes and industries. The popularity of his courses with college students suggested an eagerness to broaden research, which could lead to startling socioeconomic changes. The time seemed appropriate, then, to undertake a geological survey of the grand canal and develop an agricultural calendar, a simple tool that would have great practical value for the farmers and tradesmen who lived and worked along the canal route. The survey of Albany County was proving useful enough. Now Van Rensselaer wanted a similar survey of his namesake county, Rensselaer, which sprawled for forty-five miles down the Hudson's east shore, encompassing the area from Troy to Kinderhook. This could be the laboratory for development of the agricultural calendar.

A biweekly lecture series for the "seventy young ladies" attending Madame Emma Willard's new Troy Academy was completed in early July, 1821. On July 16, Eaton and Lewis C. Beck began to survey Rensselaer County and analyze its rocks, soils, weather patterns, and botanicals.

The field work required six weeks because, as Eaton explained in the foreword of the final report to Van Rensselaer, the focus was on "the arrangement of ecological facts, with a direct view to the improvement of agriculture. . . . My inquiries in relation to the Methods of Culture have, in this county been more particular than those made by us in Albany County. In this department, I have attempted to meet your views, by collecting materials for a kind of Agricultural Calendar, to direct the young and inexperienced farmer in regards to times of sowing, planting, harvesting &c. as well as the most approved methods for preparing his ground. Theoretical treatises on agriculture are found in abundance on every bookseller's shelves; many of which are rather calculated to perplex, than to instruct, the practical agriculturist. I have endeavored to carry your instructions into full effect, by rejecting all theory, and by preparing a concise system founded wholly on the experience of the labouring farmers of Rensselaer County. . . . I called on one, at least, in every neighborhood in all the towns; and I wrote down, in his presence, the methods of culture adopted by himself, and by his neighbors so far as had come to his knowledge. . . . I did not confine myself to crops of grass and grain, but I extended my inquiries to the subject of orchards, kitchen gardens, shrubbery, cattle, horses, sheep, swine, &c."

The system of neighborhood interviews about "Methods of Culture" and the resultant agricultural calendar of the Rensselaer survey anticipated the U.S. Department of Agriculture's county agent system by fifty years. The favorable comments the calendar won from both tenants and tradesmen persuaded Van Rensselaer to assume all of the expenses when, during the summer of 1822, Eaton presented a plan for a cross-state survey of the grand canal and recommended Professor Silliman of Yale as his principal consultant.

"The breadth of the survey is to be ten or twenty miles," Eaton explained in a September 2, 1822, letter soliciting Silliman's co-operation. "A geological map and a profile section will be published with my report. This being the greatest undertaking of the kind, I believe, hitherto known in America, I feel a deep interest in it. . . . It will probably be the most conspicuous situation that I shall ever be placed in; and the responsibility the greatest. A strip of more than three hundred miles in length, which is forever to be travelled over by the learned of all countries in stages, canals, boats, etc., I am to examine minutely and decide upon the rocks, minerals, soils and plants.

"I have now arrived at the object of this letter. It is to beg your assistance in the most difficult parts. . . . I have hitherto found much difficulty with all the secondary rocks of the west which are, geologically, above the shell-limestone. I shall be very attentive to that subject, collect large specimens in various localities, and study their relative position with minute attention. I shall enter upon that subject without any prepossessions, and cautiously reserve all opinions until I bring you a load of specimens. Now will you engage to settle that subject, if I furnish you with ample materials?"

The survey began on the morning of November 11, 1822, when Eaton and a companion rode a wagon, pulled by one horse, up the towpath past the Erie Canal locks at the Mohawk-Hudson junction. The first blizzards of winter usually roar down out of Canada and east along the Mohawk's trough during November. But the wagon continued its erratic journey from sandstone to limestone to flint to slate formations. Eaton whacked specimens from each outcropping, labeled each piece, took notes in his field book, and drove on to a farmhouse for an interview about crops.

The week before Christmas, he reluctantly turned east from the boom-town shacks of Rochester at the Falls of the Genesee. By late January, he was able to report to Professor Silliman: "I have had a geological profile drawn by one of my assistants of all the rocks from Troy to the Genesee river. . . . My forge, bellows, etc. have been set up in the front cellar, where I work when analyzing minerals."

But neither the stratigraphic system of rock classification taking shape on his sample tables nor the equally daring notion of a new type of school that would teach "the application of science to the common purposes of life" could be allowed to take too much of his time. Lectures were still his principal means of support, and in the spring of 1823, it looked as if they would be for the rest of his life. By mid-March he was dividing his energies between the basement laboratory and a series of lectures on mineralogy, zoology, and phrenology "for the accommodation of the young ladies in Troy Female Academy." Three weeks later, he crossed the snow-crested Berkshires to inaugurate a series of lectures on chemistry, geology, and zoology at "the new college at Amherst, Massachusetts."

By May he was again heading west on the grand canal's towpath to dig through Lake Erie's dolomite-over-shale dike at the new village of Lockport. For the next three months, except for the interlude of another three-weeks lecture series on "Experimental Philosophy and Chemistry" at the Troy Female

Academy, he traced glacial dumps and native strata through the maze of swamps and river valleys on the American shores of Lakes Ontario and Erie.

The strain of his rigorous schedule finally overwhelmed him in late August. He reached Troy on August 28, and though he was seriously ill for a week, he insisted on leaving on September 12 for the fall lecture series at Castleton. He came down with a form of pneumonia en route. Physicians on the Castleton faculty diagnosed his illness as "Western Lake Fever" and ordered him to bed. The fever, lung inflammation, and acute laringytis continued for a month. Eaton engaged a student, Dr. Jedediah Smith, as his assistant, then returned to Troy in late November. Dr. Smith delivered most of the Eaton lectures at Castleton, Troy, and Amherst through mid-May, 1824.

Though his voice was reduced to a rasping whisper and he was still weak from the effects of the "Western Lake Fever," Eaton worked relentlessly on the geology section of the Erie Canal survey. On February 4, he wrote Torrey that "the rock part is in the press and the engravers are at the plate. . . . I challenge all correct geologists upon the accuracy of the survey. . . . I have become acquainted with almost every individual fragment of rock, creek and hillock on the canal line."

The report, first offered for sale at Albany on April 30, 1824, was titled *Van Rensselaer's Canal Survey*. It had a revolutionary impact on geology and other emerging sciences, primarily because of the new nomenclature for native New York rocks that Eaton "proposed for examination and criticism" by fellow geologists. A century would pass before scientists would be able to view Eaton's findings objectively enough to recognize their true significance.

Eaton was still experiencing chest pains and hoarseness in March, but he left Troy for Amherst before the spring thaw and was still lecturing there when *Van Rensselaer's Canal Survey* was published. A few days after returning home, in the second week of May, he asked Van Rensselaer's permission to "deliver lectures at the villages along the canal while I am investigating the agricultural aspects for the Survey's Second Part."

His journal for the month of June, 1824 speaks plainly of his determination to popularize applied science:

June 7 (M)   Give a lecture on Natural History before the Utica Lyceum. Judge Nathan Williams present.

June 9 (W)   Give lecture this evening in Rome on Nattural History.

June 10 (Th) Return to Utica. Give a lecture at Utica Courthouse this evening at 8 o'clock P.M. introductory to a Course in Chemistry.

June 11 (F)   Lecture at Rome on Chemistry at 8 P.M.

June 12 (S)   Lecture at Utica at 6 P.M. on Botany.

June 14 (M)   Lecture at 8 P.M. on Chemistry.

June 15 (T)   Lecture at 8 A.M. on Botany. Go to Rome and lecture at 8 P.M. on Chemistry.

June 16 (W)   Lecture at 8 A.M. on Botany. Lecture this evening at Utica on Chemistry.

June 17 (Th) Lecture on Botany at 8 A.M. in Utica. Go to Rome in a wagon. Lecture this evening on Chemistry.

June 18 (F)   Lecture this morning at Rome on Botany. Lecture this evening in Utica.

June 19 (S)   Lecture at 8 A.M. in Utica on Botany—at Rome on Botany at 5 P.M.

June 21 (M)   Lecture at Rome on Galvanism.

This cultural shuttle continued for another five days, until he admitted:

June 26 (S)   Lecture at Utica on Geology. Go to Rome but am too unwell to lecture: I had been

> growing worse for eight or 10 days. I have
> a fever and inflammation in the lungs. I
> reduce for a few days. Do no business until
> Tuesday.

By July 1, he was exploring the lanes over the lovely western rim of the Mohawk valley, examining the vines and shrubs that grew so lushly beside the rattlesnake fences, chatting with farmers about crop rotation and fertilizing techniques, then rushing to reach the next lecture spot on time.

The fever was just as persistent. It finally forced him to turn east on July 20. He grumped to his journal:

> July 22 (Th)   Arrive home at 2 o'clock. I am quite un-
> well.

Amos Eaton's search for truth about the forces that had shaped the Earth and her life forms reached the threshold of a new era during the next four months. The virus that caused his "Western Lake Fever" deserves some of the credit. He had thought out the plans for the *Van Rensselaer Canal Survey* in 1821 during a period of illness in Vermont. Confined to his home by a flare-up of the same illness during the late summer of 1824, he focused his imagination on the American future.

Economic and social enrichment, he had long taught, could result from application of scientific knowledge about rocks, soils, and plants. But the few college courses, the experiments and discussions conducted by scholarly groups like the American Philosophical Society, even his own "common talk" lectures, were designed for adults and the educated elite. There should be a place where youngsters with avid curiosity could receive training *through performance* for engineering, laboratory experimentation, geological exploration, soil analysis, and other scientific professions.

Stephen Van Rensselaer was an obvious patron for so radical a school. His devotion to the popularization of the natural sciences was well known. A Rensselaer School, *if* the plan succeeded, could become the patron's most durable memorial.

Discussions about the project must have begun in the early fall of 1824. Van Rensselaer shared Eaton's enthusiasm and offered to underwrite endowments for creation of the Rensselaer School "in the building usually called Old Bank Place, at the north end of Troy."

A letter to potential trustees, mailed during the first week of November, announced the purpose of Rensselaer School as "instructing persons, who may choose to apply themselves, in the application of science to the common purposes of life."

The letter, signed by Van Rensselaer but presumably drafted by Eaton, emphasized that the "principal object is to qualify teachers for instructing the sons and daughters of farmers and mechanics, by lectures or otherwise, in the application of experimental chemistry, philosophy and natural history, to agriculture, domestic economy, the arts and manufactures."

The Rensselaer School, soon to become the Rensselaer Polytechnic Institute, was the first American institution for technical and scientific training. Like the Peale Museum, it was the offspring of one man's avid curiosity about natural history.

8 | Pupils of the Grand Canal

Nor can I doubt that [the Earth's] history throughout the long geologic ages—its strange story of successive creations, each placed in advance of that which had gone before, and its succeeding organisms, vegetable and animal, ranged according to their appearance in time, on principles which our profounder students of natural science have but of late determined—will be found in an equal degree more worthy of its Divine Author than that which would huddle the whole into a few literal days, and convert the incalculably ancient universe which we inhabit into a hastily run-up erection of yesterday.—Hugh Miller, in a lecture before the Geological Section of the British Association, Glasgow, published in his *Testimony of the Rocks*, 1857

TWO of Eaton's sons and four other Troy youngsters formed the entire student body of Rensselaer School when its "experimental term" began, on January 3, 1825. But by the following October, observers from Columbia, Union, Middlebury, Amherst, Harvard, Princeton, and Yale colleges had journeyed to Troy to study this "new technique of pedagogy."

The nucleus of the Rensselaer Plan, as historians would call it, was Eaton's dictum that "the students lecture and experiment by turns, under the immediate guidance of a professor or competent instructor. . . . At the close of the term, each student is to give sufficient tests of his skill and science before examiners. . . . The examination is not to be conducted by question and answer; but the qualifications of students are to be estimated by the facility with which they perform experiments and give the rationale."

Petitions for the establishment of similar "applied science" schools, to be funded by state governments, were circulated in Massachusetts and Pennsylvania during the winter of 1825–26. Discussions about the Rensselaer Plan brought so many pupils to the school that, by 1835, its graduates had pioneered professorships in chemistry, geology, natural history, and botany at colleges and universities in sixteen states, from Maine to Georgia and west to Louisiana.

Benjamin Silliman provided the most scholarly material about new developments in these fields. In 1818, he had purchased the seven-year-old *American Journal of Mineralogy* and redesigned it as *The American Journal of Science and Arts.* Best known as "Silliman's Journal," it was an unbiased medium for reporting experiments, theories, and discoveries in "the natural philosophies" throughout America and Europe. Typically, in 1825, Silliman printed a long letter from Peter Dobson, a Connecticut merchant. The strange scratches and polishing of rocks throughout New England, Mr. Dobson believed, could only have been caused by the movement of melting glaciers. Most scientists sighed and shook their heads after a quick scan of the letter. Only Professor Silliman would recall it a quarter-century later, when the Swiss genius, Louis Agassiz, validated and enlarged upon Peter Dobson's "hunch."

De Witt Clinton's grand canal, soon to be officially named the Erie Canal, continued to attract people eager to increase their awareness of nature. "More than half the northern immigrants arrive by way of New York, the Erie Canal, and Lake Erie," the Reverend Timothy Flint reported in his history of the West, published in 1832. "If their destination be the upper waters of the Wabash, they debark at Sandusky. The greater number make their way from the lake to the Ohio river either by the Erie & Ohio [canal] or the Dayton canal." Barge traffic was so heavy and continuous that, although shipping costs between Albany and Lake Erie dropped from $100 to $10 per ton, within eight years the barge tolls paid off the canal's construction costs of $7,143,789.

It was sound pedagogy, then, for Eaton to resurrect the field-study technique he had pioneered at Williams College and, in 1826, to organize an applied-science tour through the Erie as a postgraduate seminar for Rensselaer School's first graduating class. By using the freight boat *Lafayette*, he explained to Van Rensselaer, and installing a portable kitchen, he could hold the costs of the seven-hundred-mile journey to twenty dollars per student, and enough mineral samples and fossils could be collected and sold to museums and hobbyists to "defray the expense of the expedition."

In addition to the seven members of the graduating class, the students included Asa Fitch, a seventeen-year-old from Salem, New York; De Witt Clinton's son, George; and a twenty-seven-year-old New Yorker from Albany named Joseph Henry. Fitch, a farm youth from the Berkshire foothills, had been lured to the expedition by Eaton's speeches. He would prove so proficient that he would become a teacher at the Rensselaer School, would be appointed New York's first state entomologist, and would pioneer programs to control crop pests and diseases. George Washington Clinton, at his father's insistence, was preparing for an applied-science career. Joseph Henry, after hearing Eaton's lectures, had lost interest in his adolescent dreams of writing poems and books "like Mr. Cooper or Mr. Irving," and had recently been hired by the Albany Academy as an instructor in mathematics and natural philosophy.

Each student was instructed by Eaton to keep a daily journal of the trip and to be prepared to deliver lectures about regional geology and botany at village lyceums along the route. Since no academic credits would be given for the trip, the educational rewards of the expedition would simply be "knowing what you are looking at" in a landscape and greater fluency in public discourse. Each student lecture would be criticized by the entire group.

The trip began on the morning of May 2 as the *Lafayette* proceeded through the nine locks that circumvented the Mohawk's plunge into the Hudson. Asa Fitch began his journal with a puckish description of the *Lafayette* as "just like a coffin clapped into a canoe." The cabin, approximately fifty feet long and ten feet wide, was to be dormitory, storeroom, dining room, classroom, and library for the next nine weeks. Mattresses, commodes, dishes, food supplies, and pots were stored in cupboards. At bedtime, Fitch wrote, chairs and benches were stacked up; then the mattresses were hauled out and "arranged in two rows, which might remind one of the rows of graves in a burying ground. . . . The narrow space is so thronged that, looking down from the deck into the cabin, would bring to one's mind the Black Hole of Calcutta."

On the foredeck, ten feet beyond the cabin and precariously barricaded by a stack of chunkwood and kindling, squatted the *Lafayette*'s only source of heat: a cast-iron stove. Since the firm of Eaton, Gilbert & Co. was the principal reason Troy proudly called itself the "stove capital of America," and Eaton's brother Erasmus was the firm's senior partner, it is probable that the expedition's stove was a nickel-trimmed "potbelly" from this foundry. It required tending eight hours a day and, with the help of the Mohawk valley's prevailing western winds, fouled the cabin with smoke and cinders.

The Erie's pace was mule pace, an average of four miles an hour. The boat stopped each night at one of the "basins" where livery stables, a general store, and a tavern sprawled near the wharf. The party generally traveled less than ten miles per day, because Eaton made frequent stops to examine "the Graywacke which rests on the Transision [*sic*] argelite and extends beyond Rotterdam," or to visit places like "the polytechnic school of Dr. Yates and its interesting formations of calcerous tufa."

None of the student journals gave a detailed account of the lecture stops at Utica and Rome, but it is likely that one of them was attended by an eager young naturalist from the village of Oneida. Asa Gray, twelve years old that year, was an avid collector of unusual rocks and a devotee of the copies of "Silliman's Journal" that now and then strayed into the village. Less than a decade later, Eaton would introduce him to John Torrey, and by 1834, he was Torrey's assistant and coauthor. As professor of botany at Harvard after 1842, Dr. Gray wrote the *Manual of Botany* that became the standard reference work on American flora east of the Rockies; he also became the foremost American champion of Darwinism.

West of Lake Oneida, the canal passed the bustle and fumes of the saltworks on Lake Onondaga. Ever since the 1790's, Salina had shipped rock salt to the Pittsburgh butchers who provisioned the Ohio's "arks," "keelers," and rafts with hams, bacon sides, and jerky. The canal had enabled Salina to sell to places as far west as Michigan, Illinois, Wisconsin, and even the remote Minnesota wilderness. Now the saltworks produced that domestic necessity, baking soda, as a by-product. The Salina saltworks, the kilns in the Mohawk valley that were turning rock into cement, the iron foundries, the potteries, and the limestone quarries were Eaton's proudest examples of the practical benefits of geological awareness. The boom town developing at the east end of Lake Onondaga would be renamed Syracuse, after the ancient salt-producing center of Sicily. The *Lafayette* berthed there for three days.

The heavy gloom of Montezuma Swamp, west of Lake Onondaga, provoked long discussions about the multitude of birds, snakes, and wild grains it nourished, and its geological relationship to those mysterious round hills that thrust, like arrogant green-and-brown breasts, out of the prairie on the western horizon. That year another farm boy fascinated by "magic stones" was having visions of supernatural beings on the slope of one of these hills near Palmyra; his name was Joseph Smith. Nearby, where the Genesee River's limestone, hematite, and shale ridges had enabled Rochester to become famous as "the flour city," the Reverend William Miller similarly pondered the significance of the new way of life and wondered if the Methodism he had been preaching was prophetic enough for the times.

On hearing reports of a "burning spring" at the hamlet of Jamesport, the crew of the *Lafayette* decided to tie up there. This spring, Joseph Henry noted, was of "coal gas which burns with a dense white flame." Its existence in the same region as the "Seneca oil" seepage and almost due north of the anthracite strata of western Pennsylvania led to a discussion about the geological significance of coal deposits in New York. Could this "carburetted hydrogen," as Eaton named it, become a fuel source for industry? And what about Seneca oil? Could it be made as efficient for lamp and lantern use as whale oil?

Eaton suggested to the village wharfmaster and storekeeper

that the spring was unusual and important enough to change the village name to Gasport.

The Lockport and Niagara gorges were the great geological showcases of the expedition. Each pierced the rock ridge that formed the eastern embankment of the upper Great Lakes. The Lockport gorge, three miles long, had been blasted out to allow room for the series of locks that floated boats through the Erie-Ontario barrier. The treasure trove of marine fossils and rare minerals bared by the dig would become one of the East's richest lodes for "rock hounds" and professional researchers. "Among the various interesting minerals which are found here," George Clinton wrote, "the sulphate of strontium and anhydrous gypsum are most scarce. Ten dollars has more than once been paid for a specimen. The persons who excavated the canal did not know how high a value was attached to these productions of nature until it was more than half completed. Consequently, many of the best specimens are hidden in immense piles of stone that line the banks of the canal."

Joseph Henry also included a list of Lockport and Niagara minerals in his journal and added, "Selling of minerals has become a business of some profit . . . and large quantities of them have been sent to New York [City]. . . . A [Niagara Falls] grocer, who if he does not understand the cemical [sic] composition of minerals, at least knows full well how to sell them at a high price," offered a piece of dogtooth spar, four by two inches, for three dollars "after giving a lecture on its rairness [sic] and Beauty."

The canal's surveyors called for only a three-mile blast-out to bring boats over the 330-foot elevation between Lakes Erie and Ontario. But the Niagara River, nature's drain-off channel for the upper lakes, followed a south-to-north course thirty miles long between the new city of Buffalo and Fort Niagara's "French castle." Significant evidence of the Earth's ancient beginnings glowered from the cliffs of the seven-mile chasm that curved between Fort Niagara and the Falls. All of those seven miles, Eaton had discovered during his 1823 survey, were the same stratum as the 275-foot precipice at the Falls. A layer of limestone, eighty feet thick, lay atop an equal thickness of soft shale. The spray from the Falls, along with the annual cycle of freezing and thawing, gnawed away the underlying shale. Eventually, sections of the limestone overlay tumbled into the gorge. Then Niagara Falls moved south another few yards.

The implications of these formations were as "heretical" to most churchmen—and consequently to most politicians—as Hutton's uniformitarianism. If, as the gorge testified, the Falls had originated near the site of Fort Niagara and had since moved seven miles upstream, the Falls could be ten, twenty, even fifty thousand years old—certainly far older than they would be if one accepted the traditional interpretation of the week of Creation.

After a mineral search along the Niagara gorge, Eaton and his party toured the grain mills, rope walks, and quarries powered by the Niagara's current, and inspected the locks of the Welland Canal and the railroad line being surveyed along the old Portage Road. Back in Buffalo, the group visited the new waterfront that had been dredged and blasted from the Buffalo Creek estuary, then watched the S.S. *Superior* chuff off on her overnight run to Cleveland and Detroit.

Eastbound from Lockport, Eaton grinned at the congratulations of his students when the *Lafayette* passed beneath the new sign, *Gasport*, painted on the center span of the bridge at Jamesport.

Spectacular examples of applied science in the Genesee valley provoked another tour at Rochester. Upstream on the loamy flatlands, migrants from New England were using limestone and animal-manure fertilizers to produce the richest wheat, rye, and barley crops in America. The force of the river's 265-foot plunge through the three-mile gorge at Rochester was already being harnessed to provide the power for flour, paper, and cotton mills, as well as an iron foundry and a brewery. The stands of virgin walnut, chestnut, oak, and hemlock along the valley banks were being cut down to be used as raw materials by tanneries, boatyards, and agricultural-tool manufacturers. The energy that had been responsible for these advances was now being directed at raising funds for a canal that would parallel the Genesee to its headwaters and then veer through the Alleghenies to a junction with the Allegheny River's tumble into Pittsburgh, thus providing a floatway connection between the Hudson River, the Great Lakes, and the Ohio valley.

All of these developments were, to Eaton, a profound demonstration of the wonders that could be achieved when applied science was joined with nature. He began to wonder whether uniformitarianism, stratigraphy, and the new awareness of the Earth's age and resources might not be a form of "revelation" as indicative of God's will as any proclaimed by the traditional faiths. But the prejudices that had become a fixed part of Christian folkways forbade such realistic teaching at Rensselaer. Only research and demonstration of obvious truths could persuade the public to accept the idea of an ancient Earth. That acceptance would probably come only after centuries of bitter argument.

The chubby man who swaggered aboard the *Lafayette* on its last afternoon in Rochester and barked a request to meet "thees Professaire Eaton" had arrived at the same conclusion, and he was belligerently ready to lecture about it. He introduced himself as Professor Constantine Rafinesque, en route to Philadelphia. He wanted a ride on the *Lafayette* to Troy because he was an admirer of Eaton and needed to exchange data with him. Six years earlier, he explained, he had finally spoken out against the timidities of the curriculum at Transylvania College in Lexington, Kentucky, then resigned as the institution's professor of botany and natural history. He was thoroughly soured on teaching; he now earned a skimpy living as writer and lecturer.

Eaton was delighted. Rafinesque's *Ancient History; or, Annals of Kentucky* had been published in 1824; his letters and reviews in "Silliman's Journal" were profound, if contentious. While still at Transylvania, he had reviewed Eaton's *Index to the Geology of the Northern States* for *The American Monthly*. Eaton had been so flattered that he had memorized Rafinesque's conclusion that "when [Eaton] attempts to show that the geogony of Moses and his account of The Flood do not in the least contradict the facts which experience has revealed, when he proves that the days-of-Creation have been many periods of time, as so many learned divines have asserted and every geogonist believes, we find him engaged in a desirable act of conciliation between science and religion."

The arguments and impromptu lectures that erupted from the Eaton-Rafinesque discussions during the trip to Troy provided a momentous experience for the students. Rafinesque was, young Asa Fitch declared, "so well versed in geology and botany that we were daily going to him with specimens of plants, shells, etc. to obtain their names. . . . On rainy days, he told anecdotes or sang songs in French, Italian and English. . . . He is a universal genius, ready to investigate whatever subject presents itself to him; a full blooded polytechnic!"

Born near Constantinople, of French and German parents, Rafinesque ran away to sea and reached America in 1802, when he was nineteen. After three years of odd jobs and wandering, he returned to Europe and studied, albeit haphazardly, in France, Spain, and Germany. The theories of Cuvier, Werner, Smith, and other European naturalists drew him back to the United States in 1815. He began work on a book called *Medical Flora of the United States*—some of its conclusions would be as shocking to fellow naturalists as Smith's stratigraphy had been to the clergy. "I am convinced," he told the *Lafayette*'s explorers, "that every variety of plant is a deviation which becomes a species as soon as it is permanent by reproduction. Deviations in essential organs may then gradually become new genera." (In July of that year, 1826, Charles Darwin was seventeen years old.)

Rafinesque's verve and spirit provided a grand finale for the expedition. "I was gone only seven weeks," Asa Fitch exulted in his journal, "and yet how much I have seen. How far I have been. What new ideas I have received. How greatly my mind has improved." He enrolled at the Rensselaer School that fall and, when he was twenty-one, became an assistant professor.

Joseph Henry responded similarly to the Eaton method of "learning by doing" and became one of the examiners of the Rensselaer Plan's grading system. His association with Eaton led to a lifelong friendship with John Torrey, and he later achieved fame through his electromagnetic discoveries. In 1835, when Thomas Davenport, a Brandon, Vermont, blacksmith, invented the "Vermont electro-magnetic machine," based on Henry's discoveries, Eaton served as the intermediary between Davenport and Henry. Finally he persuaded Van Rensselaer to purchase the machine for the school, so that the students could learn firsthand about induced currents and the electromagnet's potentials not only as a source of mechanical power but also as an instrument that might be used in communication, perhaps even intercity communication.

John Torrey became a professor of chemistry and natural history at the College of New Jersey in Princeton in 1830, and he soon began urging Joseph Henry to bring his skills there too. That fall, the Reverend Joseph Powell was transferred from the Methodist chapel at Utica to the one at Palmyra. There Joseph Smith had recently published *The Book of Mormon* and founded the Church of Jesus Christ of Latter-day Saints. In Rochester, the Reverend William Miller, after carefully studying Biblical prophecies in the light of all the wickedness and lust that existed in the world, concluded that the Second Coming of Christ would occur in 1843. At Lockport, one Caleb Marsh successfully persuaded his brother-in-law, George Peabody, that a more luxurious home should be built on the Marsh farm because Peabody's sister, Mary, was pregnant again. And through the August haze and cicada drones, nineteen-year-old James Hall hiked the two hundred miles from Hingham, Massachusetts, to Rensselaer School.

The granite folds, salt hay, cranberry bogs, and fossil sharks' teeth of Boston Bay's South Shore were as intriguing to Jimmy Hall as the Berkshire and Palisade wonderlands had been to Eaton and Torrey. He discovered "Silliman's Journal" at Hingham's athenaeum while he was still in grammar school. When he was sixteen, he regularly walked into Boston in order to listen to the natural-history lectures by Silliman, Mitchill, Hitchcock, or one of the "distinguished gentlemen" from Philadelphia. The publicity Amos Eaton received in these lectures because of the Rensselaer Plan and the 1826 grand canal expedition inspired Hall's longest hike, across Massachusetts and the Berkshires to Troy. He had barely enough money to pay for his tuition and a few weeks' lodging. Unfortunately, he had no interest in, or prospects for, a profession that could earn him a living.

John Torrey, like Caspar Wistar, was a physician, as were most of the "geogony" devotees of the American Philosophical Society. Peale's Museum, although a by-product of curiosity about fossils, had been launched mainly as a showplace for the family's talents in portraiture and scenic design. Benjamin Silliman, Amos Eaton, and Joseph Henry were trained for socially acceptable, if somewhat mundane, professions before their passion for "geogony" converted them. But James Hall and Asa Fitch heralded a new era. They enrolled at Rensselaer with no aspirations other than to become researchers in and teachers of applied science.

The abundance of rocks in the Northeast had been a primary factor in bringing out the celebrated "Yankee shrewdness." The thousands of miles of stone walls, cobblestone walks and quays; the granite foundations and massive central chimneys of northern colonial and federal houses; the soapstone foot warmers, slate walks, dolomite doorsteps; all testified to the region's dependence on, and ingenuity with, rocks. James Hall grew up in an environment that encouraged this sort of ingenuity, and, in addition, he had his share of the Yankee ability to concentrate. His intensity, along with his cleverness at hunting out and trading fossils and mineral specimens during field trips, won him Eaton's friendship. Eaton arranged for Hall to do janitorial chores, cataloguing for the Troy Lyceum, and similar odd jobs so that he could earn his way through the two years at Rensselaer.

Hall's remarkable ability to locate fossils and arrange them stratigraphically was put to the test when Eaton introduced Hall to his beloved treasure troves in the Schoharie valley, particularly "the Gebhard place."

John Gebhard had been one of Eaton's earliest converts to curiosity about rocks and the mysteries they contain. The Gebhards were one of the Palatine German families that settled in the Hudson valley during 1710. After a "starving time" brought on by the neglect of the British governor, most of these Protestant refugees fled to the Mohawk and Schoharie valleys. There, through the semantic carelessness of their Scotch, Irish, and English neighbors, they were eventually labeled "Mohawk Dutch." The Gebhard farm was near Schoharie Courthouse, where ancient seas had formed the now-caverned limestone bluffs and freckled them with the shells of crustacea and crinoidea. John Gebhard and his son were assembling such an extraordinary collection of fossils and rare minerals from their fields and cliffs that Eaton considered it a worthy nucleus for a state museum. (Eventually it was purchased for the New York State Museum, and Gebhard's son became curator.)

Like Eaton, Gebhard was immediately impressed by Hall's determination to master stratigraphy and began collecting fossils for him and recommending him to other collectors. Hall became the first fossil and rock trader in New York, and probably in the Northeast. He swapped and catalogued so deftly that it provided a steady source of supplementary income to the teaching position Eaton gave him at the school.

In 1836, when the state legislature finally granted funds for a statewide geological survey, Eaton recommended that Hall be made principal geologist for one of the five districts to be studied. After the usual politicking, Hall was appointed geologist of District Four, the glacially scarred region west of Syracuse.

By 1836, the groans, hisses, and soot that accompany industrial growth had become commonplace all along the Erie Canal's route. Pennsylvania's anthracite powered the steam engines for new factories. Railroads offered the harrowing experience of a ride from Albany to Syracuse in fifteen hours. The three-ton locomotives and rebuilt stagecoaches of the Sara-

toga & Rensselaer Railroad snorted across the foot of the Rensselaer School lawn three times a day, providing a regular
source of discussion for the civil engineering students. One of
the day pupils who eavesdropped on these conversations
was ten-year-old Theodore Dehone Judah, son of the Reverend
Henry R. Judah of Saint Paul's Episcopal Church.

Down the hill at the Troy-Watervliet ferryhouse, the rough-
neck peddling papers was Charles "Bull" Crocker. The Bull's
Head Tavern on the Watervliet shore had been the birthplace of
Leland Stanford, currently odd-jobbing at his father's Elm
Grove Hotel beside the Mohawk & Hudson Railroad. Sometime
that summer, glum, lanky Collis Huntington wandered from
Connecticut to the Catskill village of Oneonta and became a
clerk at his brother's grocery store.

On his way to begin his analyses of the strata that formed the
Palmyra drumlins, the gorges of the Genesee, the goldfish-bowl
contours of the Finger Lakes, and the fossil-rich dikes of the
Niagara ridge, James Hall came within hailing distance of
Samuel B. Reed, junior engineer on the project to enlarge the
canal through Montezuma Swamp, and within a few miles of
the Geneva home of John and Daniel Casement.

As though in response to Amos Eaton's convictions, voiced in
1837, that "the rail-road is the wave of the future," the Erie
Canal and its supporting mix of applied sciences was serving as
the spawning ground for the transcontinental railroad and subsequent mechanization of the trans-Missouri West. Theodore
Dehone Judah would visualize and survey the trans-Sierra
Central Pacific Railroad, lobby for it in Congress, and persuade
Leland Stanford, "Bull" Crocker, and Collis Huntington to buy
its stock. Sam Reed and the Casement brothers would scheme
and bully Union Pacific's crews through to the Golden Spike
junction at Promontory Summit in 1869.

But, despite the abundant signs of progress along the canal
route, there were many negative reactions to geology and the
applied sciences. At Palmyra and west across the Ontario
shore, the gunfights and house-burnings that erupted against
Joseph Smith's Mormons forced the sect to flee west to Ohio.

Joseph Smith and his followers decried the "godlessness" of the geologists and their theories about an ancient Earth. The Reverend William Miller's new sect of Millerites (eventually reorganized as the Seventh-Day Adventists), who were preparing for the Second Coming, to take place in six years, held similar convictions. At Methodist, Baptist, and Lutheran academies in the Susquehanna valley and Finger Lakes region, professors continued to dramatize the lectures on natural history by displaying shell and bone fossils chiseled from local crags, and warning that they were "fashioned by the Devil to deceive you about Biblical Truth."

En route to the splendors of the Genesee's upper gorges, Hall passed the new canal town of Mount Morris. The steeple of its Methodist church was one of four that dominated the skyline of this village near the foot of the "Grand Canyon of the East." At the adjacent parsonage, the Reverend Joseph Powell wearily prepared for the transfer to his new appointment in Castile, a farming community perched on the edge of the Genesee's "high banks" ten miles south. He worried, when parish duties permitted such selfish indulgence, about the enticements of such a location for his youngest son. Some of those cliffs were three hundred feet high, and two-year-old John Wesley Powell was demonstrating fearless curiosity about pretty stones, flowers, trees, and even wasps and snakes.

A few weeks later, tracing the dolomite and shale ledges around Lockport, Hall also passed the Jabez Marsh home. If a chubby five-year-old stared through the fence palings at the intense stranger, and if the stranger acknowledged the boy, it was a charming first encounter between two who would become giants of the science that Charles Lyell would, in 1838, name *paleontology*.

9 | The Science of Early Beings

It is the customary fate of new truths to begin as heresies and to end as superstitions.—Thomas H. Huxley, *The Coming of Age of the Origin of Species* (1894)

THE three volumes covering the western section of the New York Geological Survey, each as ponderous and somberly bound as a family Bible, required five years of research, rewriting, and chart making. The volumes proved to be worth the effort, for two reasons: they accomplished what their authors expected and they advanced science. In addition, they contributed more to economic growth than the authors could have anticipated. This conjunction of scientific progress and economic growth was at the heart of the astonishing advances made in this country in the nineteenth century. Not only did scientific developments aid current business enterprises, but a growing number of people were also beginning to realize that scientific discoveries could lead to the creation of profitable industries they hadn't even been able to imagine before.

Industrial production made possible by steam-driven machines, Eli Whitney's assembly-line technique, and the canals had aroused businessmen's curiosity about the profits that might be gained through application of the findings of geological research. Respect for "the almighty dollar" outweighed the protests of fundamentalist theologians against the heresies implicit in the continued study of rock strata, fossils, and other geological materials. Consequently, appropriations for three-, four-, or five-year surveys of the mineral wealth of each state were routinely approved by the legislators; bankers and manufacturers insisted that enough "new wealth" would be discovered to justify tax boosts.

The "black dream," as some legislators called it, of the New York survey was that veins of coal would be discovered and New Yorkers would no longer have to depend on the lodes of bituminous and anthracite in Pennsylvania. New sources of granite, slate, marble, and other rocks used for construction became imperative as cities grew and buildings began to be taller than two or three-stories. The transition from fireplace to stove in most homes, the railroaders' adoption of the new T rail and flanged car wheel, and the advent of cast-iron architecture all underscored the necessity to discover more iron deposits. There was even a flurry of interest in the petroleum oozing from springs and ponds along the Pennsylvania border, although it seemed certain that the patent-medicine peddlers would use all of the available supply for their "snake oil" and "magick liniments." In any event, whale oil and tallow, used as lamp fuel, seemed to be too plentiful for petroleum to be a serious competitor.

The business community's approval of geological research encouraged the introduction of geology to the natural-science curriculums of most colleges and universities. In addition, adult education became more popular than ever, and Amos Eaton's "simplification" technique became the most widely used method of instruction in the educational associations known as lyceums. The lyceum system of subscription memberships to a series of "self-improvement" lectures was initiated in 1826 by Josiah Holbrook, an alumnus of Benjamin Silliman's lectures. A decade later the invalid son of New England's first cotton-mill millionaire, Francis Cabot Lowell, endowed the Lowell Institute in Boston as a center for self-improvement lectures and projects; Silliman delivered the inaugural series of lectures. By that time, hundreds of lyceums were being offered in athenaeums and church parlors, with New York City's Cooper Union and the summer programs at Chautauqua, New York, eventually serving as models.

At the same time, nearly every major city and county seat, it seemed, was anxious to have its own version of the Peale Museum. In Bridgeport and New York City, a man who owed much to the Peales, Phineas T. Barnum, assembled freaks, oddities, and gimcracks for the 1842 opening of his American Museum.

In Philadelphia, devotees of geology, botany, and zoology had founded the Academy of Natural Sciences, which worked so closely and harmoniously with the American Philosophical Society that, in 1849, it would take over the Society's collection of fossils and some of the exhibits from the Peale Museum.

In 1840, Silliman, Torrey, Henry, and Eaton helped found the Association of American Geologists, which, two years later, was expanded to the Association of American Geologists and Naturalists.

William C. Redfield of Connecticut, an amateur weather forecaster and the inventor of a "safety barge" for cabin passengers that could be towed behind steamboats, was instrumental in founding the American Association for the Advancement of Science in 1848.

While the gold rush wagon trains moved westward, and Congress continued the grim task of balancing free territory with slave territory, these natural-science professors and prosperous hobbyists set the pattern for the educators' convention: a week at a resort hotel; a daily agenda of several hour-long, elaborately documented and usually abstruse "papers," read in a monotone; discreet job hunting at evening receptions; sitting stiffly in a "boiled shirt" through the annual banquet and president's address; the morning-after hang-over; and an outbreak of venereal "pox" two weeks later among some of those who had "gone on the town."

In the half century since Christian Michaelis's *Mammut* teeth had inspired Charles Willson Peale to found the first effective American museum, fossils and the search for truth about the Earth's age had been the stimuli for:

- the emergence of science as a socioeconomic force in America;

- the acceptance of geology, natural history, civil engineering, and applied science as valid professions;
- the switch from wood to fossil plant life as a source of heat and power;
- the development of adult education programs that provided not only intellectual enrichment, but also amusement and emotional stimulation;
- the birth of scientific associations and their rigidly dull "agendas."

And now, while James Hall probed new evidence of an ancient Earth from the fossil strata of the Ontario and Erie shores, the job of identifying and classifying fossils was finally being given a professional designation in England.

At the same time that Amos Eaton was beginning his studies at Yale, Charles Lyell was traveling from his family's estate at Forfarshire, Scotland, to Oxford in order to begin preparing for a law career. Because he had heard a great deal about him, Lyell enrolled in a course offered by the Reverend William Buckland, a young reader in mineralogy who was putting so much zest into his lectures that, rumor had it, the rheumy deans would soon be forced to acknowledge geology as a science worthy of inclusion in the curriculum. Lyell found Buckland's lectures so inspiring that he abandoned law in favor of geology as his career goal, despite the vehement disapproval of his father. By 1823, Lyell, as secretary of the Geological Society of London, was exploring the strata of Scotland, England, and France with Buckland.

A fossil search through France and Italy during 1827–28 with Mr. and Mrs. Roderick Murchison became a significant journey in the evolution of geology into a "practical" science. Murchison had been an army officer and gentleman farmer until his wife's enthusiasm about sketching rock strata and fossils persuaded him to study with Buckland at Oxford and with the Reverend Adam Sedgwick, a lecturer in geology, at Cambridge. The year-long carriage tour with the Murchisons not only convinced Lyell to complete the textbook he had started, called *Principles of Geology*, but also provided data that enabled him to identify the four subdivisions of the Earth's most recent rock strata collectively known as the *Tertiary* era.

In collaboration with a French conchologist, Gerard Deshayes, Lyell examined more than forty thousand fossils he had collected between Scotland and Sicily, compared them with species still living, and then classified the rock strata chronologically "by reference to the comparative proportion of living species of shells found fossil in each." He identified the four epochs within the Tertiary era as: Newer Pliocene, Older Pliocene, Miocene, and Eocene, and insisted that evidence of human habitation was plentiful only in the Newer Pliocene.

Although he did not attempt to interpret the Tertiary strata in terms of years, it was obvious that each subdivision covered thousands of centuries. Thus devotees of geology could speak of the Earth's age in terms of "millions" rather than the traditional Judeo-Christian "thousands." This conclusion, along with his use of Smith's stratigraphic system as the guide for identifying and chronologically arranging each rock layer, established Lyell as the outstanding champion of Hutton's uniformitarianism.

The three volumes of his *Principles of Geology*, published between 1829 and 1833, became the essential textbook of the profession. In its discussion of fossils and the vital role they had played in the development of geology as an exact science, Lyell urged adoption of a Greek-rooted word, meaning "the science of early beings," as the professional name for research of the types of plant and animal fossils embedded in the Earth's layered crust. His suggestion was adopted throughout Europe and the Americas. Thus, three centuries after the heretic curiosities of Leonardo da Vinci, the dawnseekers' science was given the name of *paleontology*.

Lyell's success in systematizing rock layers encouraged the Murchisons and the Reverend Sedgwick to explore the strata of fossil-bearing rocks exposed in sections of Wales. Murchison completed enough of his surveys before July, 1835, to write a magazine article announcing his discovery of another geologic period. He called it the *Silurian*, after the Silures, a pre-

Celtic tribe that once dominated what is now southern Wales. The Reverend Sedgwick maintained that the stratal sequence he was exploring contained fewer fossils than the Silurian strata, and so must be more ancient. He named it the Cambrian system, after the Roman name for Wales.

Despite heated arguments about the "overlap" of their claims, the army veteran and the cleric remained fast friends, and from 1836 to 1839 they worked together to determine the stratigraphic relationships of a third rock system in Devon that was eventually accepted as the Devonian system.

Reports in both British and American scientific journals about the Lyell, Murchison, and Sedgwick systems caused James Hall to be ultracautious in his examinations and analysis of the rock strata of western New York. The banks of the Erie Canal at Lockport were a rich source of trilobites and fish and other fossils. Working upstream from Rochester to the Genesee headwaters in the Allegheny wilderness, he concluded that the exposed strata almost perfectly matched the fossil sequences that Murchison and Sedgwick were identifying as Devonian. "It was the Genesee valley," Hall's biographer, John M. Clarke, confirmed eighty years later, "with its great falls at Rochester, at Portage, its impressive gorge between the 'High Banks' and its complete trans-section of the State from Pennsylvania to Lake Ontario that unlocked the geological history of western New York. It remains today, and must ever remain, not only a monument to Hall's transcription of the record but the wide-open leaves of the great Devonian book."

Puzzling over and through the fossil-studded hills east of the Genesee and probing the depths of the Finger Lakes, Hall worked his way toward another discovery that fellow geologists would consider too radical for the next fifty years. Mountain chains, he concluded, must be products of a complex process that had taken millenniums to complete. First there was a huge trough, formed by a great downward thrust—eventually named the *geosyncline*—of subterranean rocks or liquids. This trough must have eventually filled with sediment created by earthquakes, seasonal erosions, rivers, and intermittent submergence

in an ocean. Some of the sediment accumulated to heights of
eight or ten miles before pressures and natural forces transformed it into rock. Then, Hall reasoned, this newly made rock
was slowly forced up into mountain ranges by pressures from
the underlying mantle, by earthquakes, and by volcanic action.
This balancing and counterbalancing behavior of the Earth's
interior, called *isostasy*, was not accepted until the 1890's. The
force patterns of isostasy remain a mystery in the 1970's, partly
because exploration of "outer space" has taken precedence
over research of the Earth's "inner space."

Throughout his investigations, Hall retained his enthusiasm
for collecting fossils. On July 22, 1839, he wrote Benjamin
Silliman to offer a "valuable collection . . . of Geological
specimens, fossils, etc. which the State of New York will furnish . . . an interesting and instructive suite, being the exact
counterpart of the groups of rocks of the Silurean [*sic*] System
as described by Mr. Murchison.

"We have besides this advantage," he continued, "that our
rocks are better developed, have been subjected to less dis-
turbance and contain a much greater number of fossils. I think
we shall find nearly every one figured by Mr. M., besides many
others. I have lately found the new genus of Trilobite which he
figures, the *Bamastis* [*sic; Bumastus*] and in some examinations within a few weeks have discovered the remains of fossil
fish in the Old Red sandstone, which is an interesting and important member of the family, though it has usually been over-
looked. And I am not aware that the remains I speak of have
been before noticed. They are the scales, jaw and fin of a large
fish, perhaps the *Gyrolepis*. The scales are an inch or more in
length and nearly as broad.

"While engaged in the Geological Survey I do not like to
engage to furnish rocks and fossils very extensively as it might
be thought to interfere with my duties to the state, but I have
many on hand which I have been collecting for several years
past. It is also my intention as soon as the series is completed to
engage extensively in collecting the rocks, fossils and minerals
to illustrate the Geology of New York in the most perfect

manner. For this enterprise I shall be glad of your countenance and aid, but at present I do not wish to make the matter public. . . . If the price you pay will warrant it, I shall be glad to supply those of the Tertiary. The Primitive rocks of this state, as well as the minerals, can be readily supplied, principally from what I now have on hand. The rocks and fossils of the Silurean [*sic*] System cannot be collected properly except by someone acquainted with the whole series from minute examination."

Silliman's offer for the "New York geological specimens" must have been satisfactory; the following January, Hall wrote Benjamin Silliman, Jr., that "I am greatly obliged to your Father and to yourself for the favorable manner in which you have your desire to forward my object in regard to the geological collections." Hall became a contributor to "Silliman's Journal" and began corresponding with geologists in Germany, Great Britain, and Switzerland. His most loquacious correspondent was Jean Louis Rodolphe Agassiz, professor of natural history at the University of Neuchâtel and now completing his five-volume *Recherches sur les poissons fossiles.*

Hall was involved in the tedious chores of proofreading and dummying the volumes of the state survey when he received a letter from John Torrey advising him that Charles Lyell was scheduled to "reach Boston by the first week in August" (1841) to begin "a tour of some months in this country in exploring some of the more interesting geological regions. . . . He wishes to devote eight to ten weeks in exploring the geology of the State of New York, and the country about the Falls of Niagara and Lake Ontario. He wishes to have your company and to know where you may be found soon after his arrival. He desires me to inform you that he has just returned from a geological tour in the border country of England and Wales where he has been examining the older or Silurean [*sic*] strata with a view of comparing them with those of the United States."

Lyell and his wife, Mary, reached Albany on August 16. "At Albany," Lyell wrote in his *Travels in North America*, "I found several geologists employed in the Government survey,

and busily engaged in forming a fine museum to illustrate the
organic remains and mineral products of the country. . . . The
legislature four years ago voted a considerable sum of money,
more than 200,000 dollars or 40,000 guineas, for exploring its
Natural History and mineral structure, and at the end of the
first two years several of the geological surveyors, of whom four
principal ones were appointed, reported among other results,
their opinion that no coal would ever be discovered in their
respective districts. . . . Accordingly during my tour, I heard
frequent complaints that, not satisfied with their inability to
find coal themselves, the surveyors had decided that no one else
would ever be able to detect any, having had the presumption to
pass a sentence of future sterility on the whole land. Yet, in
spite of these expressions of ill-humour, it was satisfactory to
observe that the rashness of private speculators had received a
wholesome check; and large sums of money, which for twenty
years previously had been annually squandered in trials for
coal in rocks below the carboniferous series, were henceforth
saved to the public."

The sunrise was gilding Stephen Van Rensselaer's sheep
pasture on August 20 when James Hall escorted the Lyells and
their luggage to the Hudson & Mohawk Railroad terminal to
begin the exploration of "the entire succession of mineral
groups from the lowest Silurean [*sic*] up to the coal of Pennsylvania." Once over the valley ridge, the train clattered across
the sand dunes to the valley of the Mohawk at Schenectady,
then west along the Great Lakes runoff gouge. The party
stopped "here and there to examine quarries of limestone," and
made "a short detour through the beautiful valley of Cedarville
in Herkimer County."

"The excavations also made for the grand canal at Lockport," Lyell wrote, "afforded us a fine opportunity of seeing
these older fossiliferous rocks laid open to view. . . . In the
course of this short tour, I became convinced that we must turn
to the *New World* if we wish to see in perfection the oldest
monuments of earth's history, so far at least as it relates to its
earliest inhabitants. Certainly in no other country are these

ancient strata developed on a grander scale, or more plentifully charged with fossils; and, as they are nearly horizontal, the order of their relative position is always clear and unequivocal. They exhibit, moreover, in their range from the Hudson River to the Niagara some fine examples of the gradual manner in which certain sets of strata thin out when followed for hundreds of miles, while others previously wanting become intercalated in the series. . . . Another interesting fact may be noticed as the result even of a cursory survey of the fossils of these North American rocks, namely that while some of the species agree, the majority of them are not identical with those found in strata, which are their equivalent in age and position, on the other side of the Atlantic. . . . It has usually been affirmed that in the rocks older than the carboniferous, the fossil fauna in different parts of the globe was almost everywhere the same; but judging from the first assemblage of organic remains which I have seen here, it appears to me, that however close the general analogy of forms may be, there is evidence of the same law of variation in space as now prevails in the living creation."

Lyell's desire to editorialize about American fossils must have been responsible for the brevity of his description of the visit to the Lockport canal gorge. Hall took him there because this was the richest source he knew of for sulphate of strontium, anhydrous gypsum, the rare *Bamastis* [*sic; Bumastus*] trilobite, and the fossils of the big-scaled fish he had written Professor Agassiz about. Here could have occurred a meeting between Lyell and paleontology's most famous popularizer.

The person who might have brought them together would have been Colonel Ezekial Jewett, whom Lyell certainly met either at the Lockport gorge or a day or two later at Fort Niagara. A fifty-year-old soldier who had served in the War of 1812 on the Niagara frontier, then gone off to Chile to serve under Simon Bolivar, Jewett in 1826 became commandant of the semideserted Fort Niagara. The routine of spying on the British garrison at Fort George, across the river, and being spied on by them, was a bore. The 1826 visit by Amos Eaton and his students sparked Jewett's interest, and he began reading

books about geology and collecting fossils. He grew so proficient that, in 1856, Hall would sponsor Jewett for the post of
curator at the state museum.

The Lockport gorge was Jewett's favorite hunting place. During the spring of 1841, he struck up an acquaintance with a
chubby, brown-haired boy who was trying to pry a fossil fish
out of the shale with a rusty hoe blade. Jewett extracted the fish,
identified its genus, then delightedly answered the boy's bar-
rage of questions about how and why "that old fish" got trapped
in a rock this far out of the water. Thereafter, the colonel and
the ten-year-old became rock-hunting companions. After a time,
Jewett learned that "Othy," as the boy called himself, was
Othniel, the son of a Chestnut Ridge farmer named Caleb
Marsh. His mother had died when Othniel was three years old.
His uncle, George Peabody, lived across the ocean in a great
city called London, and was supposed to be very wealthy.
Othniel's father did not like him to play in the gorge, let alone
bring home a lot of rocks. Whenever Colonel Jewett wanted
Othniel to join him on a fossil dig, it would be best to ride past
the Marsh farm and whistle "Yankee Doodle." Then Othniel
would sneak over as soon as he could. He was saving the stones
and fossils the colonel had given him; they were hidden in a
bucket buried behind the third row of hollyhocks on the north
side of the outhouse.

After examining the sand ridge that parallels Lake Ontario's
shore line at distances varying from three to eight miles and
concluding that it was a remnant of an ancient lake, Lyell and
Hall reached Niagara Falls on August 27. A five-day exploration of the gorge convinced Lyell, as Eaton and Hall had previously been convinced, that the falls had originated at Lewis-
ton, then gouged out the seven-mile gorge north of their 1841
location. The wild majesty of the falls thrilled Lyell, but he
would not venture a guess about their age, because "however
much we may enlarge our ideas of the time which has elapsed
since the Niagara first began to drain the waters of the upper
lakes, we have seen that this period was only one of a series, all

belonging to the present zoological epoch; or that in which the living testaceous fauna, whether freshwater or marine, had already come into being.

"If such events can take place," he wrote, "while the zoology of the earth remains stationary and unaltered, what ages may not be comprehended in those successive Tertiary periods during which the Flora and Fauna of the globe have been almost entirely changed. Yet how subordinate a place in the long calendar of geologic chronology do the successive Tertiary periods themselves occupy? How much more enormous a duration must we assign to many antecedent revolutions of the earth and its inhabitants?

"No analogy can be found in the natural world to the immense scale of these divisions of past time, unless we contemplate the celestial spaces which have been measured by the astronomers. Some of the nearest of these within the solar system, as for example the orbits of the planets, are reckoned by hundreds of millions of miles, which the imagination in vain endeavors to grasp. Yet one of these spaces, such as the diameter of earth's orbit, is regarded as a mere unit, a mere infinitesimal fraction of the distance which separates our sun from the nearest star. By pursuing still further the same investigations, we learn that there are luminous clouds, scarcely distinguishable by the naked eye, but resolvable by the telescope into clusters of stars, which are so much more remote, that the interval between our sun and Sirius may be but a fraction of this larger distance.

"To regions of space of this higher order in point of magnitude, we may probably compare such an interval of time as that which divides the human epoch from the origin of the coraline limestone over which the Niagara is precipitated at the Falls. Many have been the successive revolutions in organic life, and many the vicissitudes in the physical geography of the globe, and often the sea has been converted into land, and land into sea, since that rock was formed. The Alps, the Pyrenees, the Himalayas have not only begun to exist as lofty mountain

chains, but the solid materials of which they are composed have
been slowly elaborated beneath the sea within the stupendous
interval of ages here alluded to."

The revolutionary significance of this reverie by the forty-
four-year-old Scotsman as he stared into the rainbowed spray of
Niagara can be understood only when it is viewed in historical
perspective. When Lyell was born, any expression of doubt
about the "sacred truth" of the traditional interpretation of
Genesis was punishable in the new United States by a three-
year jail sentence and virtual loss of civil rights. Then Thomas
Jefferson knew that as a public servant he had to insist that the
Almighty would not permit any species to become extinct, and
congregations nodded grim agreement to the cry of "heresy"
thundered from the pulpits against Hutton and his uniformitarianism. Lyell was a schoolboy when Benjamin Silliman's
lectures demonstrated to James Fenimore Cooper, Washington
Irving, Amos Eaton, and others that the "giant footprints"
found in New England's rocks were records of the chaos of
Noah's Flood and upheld Ussher's edict that the Creation had
taken place in 4004 B.C. William Smith's stratigraphy began to
gain credence while Lyell was at Oxford.

Now, at forty-four, staring at the mightiest cataract on the
ancient land mass curiously called the New World, he could
conceive of and freely muse about an Earth so ancient that "the
imagination in vain endeavors to grasp it." No passage in nineteenth-century literature so succinctly reveals the rapidity with
which science, and its technological offspring, developed and
was accepted as respectable by civilized men on both sides of
the Atlantic, despite the campaign waged against it by religious
leaders.

In 1841, Benjamin Silliman, Amos Eaton, Charles Willson
Peale, Cuvier, Jefferson, and Werner were philosophically as
out of date as Francis Bacon and Walter Raleigh. Awareness
had swung full circle back to the Brahman and Aristotelian
belief in an incomprehensible infinity. Dawnseeking now depended on solemn-eyed youngsters like Othniel Marsh and
Lyell's introspective young friend, Charles Darwin.

From the Niagara frontier, Hall led the Lyells back over the ridge of the upper lakes and southeast across the Silurian lime-stone, marls, and gypsum to the upper gorge of the Genesee between Portage and Castile. At Castile, one of the youngsters who watched the strangely clothed "Britisher" clambering along the Silurian precipice could have been young John Wesley Powell, aged seven. He may even have confided to Mary Lyell, placidly waiting in a gig, that within a few weeks he was moving "way far out west to a place called Ohio." Or perhaps, since truth frequently *is* stranger than fiction, the future was twice met by Lyell in the Lockport gorge, with Othniel Marsh and Colonel Jewett rockhunting on the bank and Johnny Powell staring over the railing of a packet boat making its way toward the steamship docks at Buffalo.

After a successful dig for mastodon bones in the prairie bog near Geneseo, and satisfaction over the discovery that the giant had died "in the shell marl below the peat, and therefore agreed in situation with the large fossil elks of Ireland," the Lyells and Hall parted company. Hall hurried back to his proofreading duties at Albany, and the Lyells went south into Pennsylvania to examine carboniferous strata and Blossberg coal mines before returning to Troy to pay their respects to the bedridden Amos Eaton.

"The mind of this pioneer in American geology was still in full activity, and his zeal unabated," Lyell noted after his September 20 visit with Eaton. "His *Survey of the Erie Canal* was the earliest account of the Niagara district, but nearly all of the rocks and groupings [he named] have been since adopted by the New York surveyors."

Amos Eaton was on his deathbed. Charles Willson Peale had died in 1827, and scientific awareness had since gone so far beyond the concepts of his museum that the Wallkill *Mammut* and much of its other exhibits were about to be sold to P. T. Barnum. Georges Cuvier had been dead for a decade.

On September 28, the Lyells admired "the cleanliness and beautiful avenues of various kinds of trees" in Philadelphia, while resting before "the examination of the cretacious [*sic*] strata of New Jersey." Eager to pursue his interest in fossils, Joseph Leidy was beginning classes at the University of Pennsylvania's medical school. Across the Alleghenies, John Wesley Powell and Ferdinand V. Hayden perfected their natural-history skills under the tutelage of farmer and trapper hobbyists. Colonel Jewett continued to shape Othniel Marsh's career at the Lockport gorge.

Now that paleontology finally had a name, it was time for the "dragon hunters."

10 | The Dragon Hunters

The combination of such characters, some as the sacral ones, although peculiar among Reptiles, others borrowed as it were from groups now distinct from each other, and all manifested by creatures far surpassing in size the largest of existing reptiles, will, it is presumed, be deemed sufficient ground for establishing a distinct tribe or suborder of Saurian Reptiles, for which I would propose the name of Dinosauria.—Richard Owen, in a lecture before the British Association for the Advancement of Science, Plymouth, 1841

A month before James Hall guided the Lyells across the Silurian wonderlands of western New York, the members of the British Association for the Advancement of Science traveled through the Devon meadows toward their annual meeting at the port of Plymouth.

One of the most intriguing matters scheduled to be presented to this array of physicians, teachers, museum curators, chemists, and wealthy amateurs was the report on the British fossil reptiles. Folk tales about gigantic fanged dragons that spouted fire and had an impenetrable hide were as ancient, and as universal, as the story of the Great Flood. The first hint that Chinese, Scandinavian, and English dragon lore might have been based on fact came in an article written by Lyell's mentor, William Buckland, in 1824 for the *Transactions of the Geological Society of London*. Buckland's curiosity had been aroused by the strange jawbones and teeth in a block of slate on display at the Oxford Museum. The block came from a Mesozoic quarry at Stonesfield. The jaw was shaped in an elongated U, like an alligator's. The teeth looked like miniature daggers and were deeply set in the sockets. Buckland and another avid geologist, the Reverend W. D. Conybeare, decided that this was part of the skeleton of a huge lizard. Subsequent discovery of chunks of pelvis, vertebrae, and leg bones convinced them that this prehistoric dragon had had "a length exceeding forty feet and a bulk equal to that of an elephant seven feet high." They named the creature *Megalosaurus*, and Buckland titled his paper "Notice on the *Megalosaurus* or Great Fossil Lizard of Stonesfield."

Meanwhile, the vigilance of Mary Ann Mantell launched a second dragon quest. Mrs. Mantell often accompanied her physician husband on his rounds in the South Downs, near their home at Lewes, in Sussex. Dr. Gideon Mantell, a close friend of Charles Lyell, was an enthusiastic paleontologist and was then completing the manuscript of a book, *The Fossils of South Downs*. Mary Ann Mantell, despite the demands of rearing their four children, had done the 364 detailed drawings of South Downs fossils that would illustrate the volume.

On a spring afternoon in 1822, Mrs. Mantell was especially alert as she strolled down a lane while her husband visited a patient. Thus she did not fail to see, high on the bank, a rock studded with scissor-blade teeth. She was waiting excitedly for help to pry it out when her husband returned to the buggy.

Inspection of the rock and the teeth initiated a three-year search. The dentures evidently belonged to "an herbivorous animal," Mantell recalled years later, "and so entirely resembled in form the corresponding part of an incisor of a large pachyderm ground down by use, that I was much embarrassed to account for its presence in such ancient strata; in which, according to all geological experience, no fossil remains of mammalia would ever be discovered; and as no known existing reptiles are capable of masticating their food, I could not venture to assign the [teeth] in question to a Saurian."

Buckland and Conybeare examined the rock one afternoon at a meeting of the Geological Society and agreed that the teeth belonged to a large fish and that Dr. Mantell must have erred in the identification of the strata. Charles Lyell was more open-minded—he offered to take the rock to Paris for analysis by Cuvier; he returned the rock months later with the report that

Cuvier had tersely dismissed the teeth as the "upper incisors of a rhinoceros."

But Lyell, as dubious as the Mantells about both the Buckland-Conybeare and Cuvier conclusions, urged a search through the collection of anatomical specimens at the Hunterian Museum of the Royal College of Surgeons. There another researcher, recently returned from Central America, remarked on the teeth's resemblance to those of the iguana lizard, although they were far larger than those of a contemporary iguana adult. A detailed comparison of the fossils with the iguana teeth in the Hunterian storage bins indicated that the fossils must be remnants of a giant prehistoric lizard. Mantell reported his conclusions in a paper titled "Notice on the *Iguanodon,* a newly discovered Fossil Reptile from the Sandstone of Tilgate Forest in Sussex." The paper was published in 1825 in the *Philosophical Transactions of the Royal Society,* and Cuvier was one of the first to concede: "Do we not have here a new animal, an herbivorous reptile? And even among the modern terrestrial mammals it is within the herbivores that one finds the species of the largest size, so among the reptiles of another time, then the only terrestrial animals, were not the greatest of them nourished on vegetables? Time will confirm or reject this idea. If teeth adhering to a jaw would be found, the problem might then be resolved."

Cuvier's hope was realized in Februray, 1834, when a routine blast at a quarry near Maidstone, thirty miles northeast of Lewes, in Kent, shattered the skeleton of an "animal of great magnitude." The quarry owner, W. H. Bensted, was an amateur geologist, so the fragments were assembled and glued and eventually sold for twenty-five pounds to Dr. Mantell. There were enough vertebrae and leg, rib, and tooth fragments to enable a guess at both the size and classification of the lizard they were now calling *Iguanodon.*

Naturally, the publication of Buckland's and Mantell's descriptions of the *Megalosaurus* and *Iguanodon* fossils stimulated exploration of the Jurassic, Cretaceous, and other ancient rock strata of Great Britain and Europe. By 1841, enough bone and tooth chunks had been chiseled out to establish claims for nine different types of the monsters. The implications of these discoveries created new doubts about the "morality" of science and refueled the protests of thousands of conventional clergymen and their flocks. Geologists, by and large, now agreed that some of the rock strata were millions of years old; here was substantial evidence that families of dragons had stalked an already ancient Earth thousands of centuries before 4004 B.C.

Did man also exist then? If not, how and why had the legends about dragons become so fixed and so widespread in man's folklore? Where—and perhaps even what—was man when *Megalosaurus* and *Iguanodon* foraged for food to fill their massive bellies?

Buckland had already thought out an explanation to pacify the defenders of recent Creation and, for the time being, to lull their outrage against "ungodly science." The assumption must be, he concluded, that Moses and the other recorders of Genesis were concerned only with the creation of God's most precious achievement: the human being made in His image. "Though Moses confines the details of his history to the preparation of this globe for the reception of the human race," he explained to his students at Oxford, "he does not deny the prior existence of another system of things, of which it was quite foreign to his purpose to make mention, as having no reference to the destiny or to the moral condition of created man."

But the questions about dragon fossils still hung in the air during the dusty ride to Plymouth. Certainly, the professors, curators, and clerics agreed, there could be no more competent authority to diagnose both the physical and the topographic evidence than the hawk-nosed, cadaverous Richard Owen, considered by many to be the best anatomist in Britain.

Born at Lancaster in 1804 and apprenticed to a surgeon when he was sixteen, Owen's skills as an anatomist had won him the appointment as assistant to the conservator of the Hunterian Museum when he was twenty-two. The Buckland and Mantell discoveries so intrigued him that he joined the great dragon search. He had identified and named two of the nine known families of the giant lizards. Thus, despite the demands of his new post as Hunterian Professor, he undertook an intensive analysis of all the giant lizard bones and the strata from which they had been so carelessly chiseled and blasted. His paper before the Association for the Advancement of Science, it was said, would not only detail his findings but would also bluntly state if these were the remains of a monstrous race. Once again, man's knowledge about the earth's beginnings might undergo radical changes.

"Was this," the professors and clerics mused over their evening tankards, "the serpent that proffered the apple to our collective ancestress?" Certainly Dr. Owen would never dare to question the Bible's dictum that man had been created in God's image, without any preliminary "experimentation." There had been long discussions among the directors of the Association about the scathing attacks being hurtled from the pulpit of York Minster by Dean Cockburn. A pamphlet by Cockburn accused the Association of heresy and immorality; it had sold out five printings within two years. Both personal conviction and political expediency would convince Owen to avoid any inferences about man in his analysis of those bones. It was bad enough to refute the assumption that the Creation required only 144 hours.

The eagerly awaited speech was a model of the polysyllabic jargon, long sentences, droning detail, and evasiveness that would become the trademarks of the scientific paper. Its marrow was in the sixty-eight-word sentence quoted at the beginning of this chapter. In it, Owen conceded "sufficient ground for establishing a distinct tribe or suborder of Saurian Reptiles" and proposed the coined Greek compound *Dinosauria* ("terrible lizard") as a name. That sentence was as momentous as any composed during the first half of the nineteenth century, although subsequent research proved that the huge beasts of Earth's Mesozoic age were not lizards, nor were all of them "terrible." *Dinosaur* became a household word—children fantasized about the monsters, and interest in museums was renewed when dinosaur bones were added to their collections. The "terrible lizard" proved to be the most effective popularizer of natural history.

The outrage against Owen's revelations echoing from pulpits and vestries that fall was only a small warning of what was in store. In eastern Europe, the press, if it knew about Owen's paper, discreetly avoided mentioning it. In western Europe, most Christian denominations continued to be wary of natural science as a "potential hell raiser." In Great Britain, however, Church of England and Methodist clerics permitted references to dinosaurs and Reverend Buckland's reinterpretation of Genesis in editions of church-endowed encyclopedias and histories revised during the 1850's.

Pope Pius IX forbade the scientific congress of Italy to meet in 1850 at Bologna. Textbooks written in France as late as 1877 continued to insist that "fossils derived from Noah's Flood." In Lithuania, in 1869, Archbishop Macarius of the Eastern Orthodox Church preached that "Creation in six days of ordinary time and the Deluge of Noah are the only causes that geology seeks to explain." Throughout Germany, while congregations steadily dwindled, Lutheran ministers continued to preach and write that "geology is rendered futile and its explanations vain by two great facts: the Curse which drove Adam and Eve out of Eden, and the Flood that destroyed all living things save Noah, his family, and the animals in the Ark."

In America, the Unitarians and Quakers had taken a positive approach to natural-history research since the early years of Franklin, Jefferson, Peale, and Wistar. But the Lutherans, Roman Catholics, and Baptists, like the new Mormon and Millerite sects, still clung tenaciously to the concept of recent Creation.

At Yale, the Sillimans continued to publish all views and report on all new research in the natural sciences, but they themselves remained faithful to the theory of the Flood and the recent Creation of man. Amos Eaton's old friend, the Reverend Edward B. Hitchcock of Amherst College, became absorbed in speculations about the footprints in the Triassic sandstone of the lower Connecticut valley. Although he disagreed with the Congregationalist authority who, about 1800, identified these as "footprints of Noah's raven," he was convinced that they had been made by "gigantic birds." He secured an endowment from Samuel Appleton of Boston to build an "Appleton Cabinet" on the Amherst campus, to be filled with slabs containing the three-toed tracks of "pachydactylous, the thick-toed birds" and "leptodactylous, the narrow-toed birds."

"It is no idle boast to say," Hitchcock wrote in 1848, "that I have devoted much time and labor and thought to these memen-toes of the races that, in the dawn of animal existence in the Connecticut valley, tenanted the shores of its rivers and estu-aries. Whatever doubt we may entertain as to the exact place on the zoological scale which these animals occupied, one feels sure that many of them were peculiar and gigantic; and I have experienced all of the excitement of romance as I have gone back into those immensely remote ages and watched those shores along which these enormous and heteroclitic beings walked. Now I have seen, in scientific vision, an apterous bird, some twelve or fifteen feet high—nay, large flocks of them—walking over the muddy surface, followed by many others of analogous character, but of smaller size. Next comes a biped animal, a bird, perhaps, with a foot and heel nearly two feet long. Then a host of lesser bipeds, formed on the same general type; and among them several quadrupeds with disproportioned feet, yet many of them stilted high, while others were crawling along the surface with sprawling limbs."

Actually, Hitchcock's guesses, although based totally on the sandstone footprints, were as valid as many of Richard Owen's conclusions, based on the jumble of bones in the Oxford and Hunterian collections. Subsequent research would indicate that modern birds and lizards are both offshoots of the family tree of the Triassic reptiles known as thecodonts.

In 1853 and 1854, Owen collaborated with a popular British artist, Waterhouse Hawkins, on developing the first life-sized statues of dinosaurs. The Crystal Palace, built for the Interna-tional Exposition of 1851, had been so popular that authorities decided to make it a permanent center for fairs and exhibits. Owen and Hawkins were commissioned to develop a dinosaur family group on one of the plazas of the new Crystal Palace grounds. Hawkins set up a workshop in the Palace and, with Owen meticulously going over each detail, produced an array of nightmarish beasts that, half a century later, would cause Othniel Marsh to growl, "So far as I can judge, there is nothing like unto them in the heavens, or on the earth, or in the waters under the earth. . . . The dinosaurs seem to have suffered much from both their enemies and their friends." Dr. Mantell's *Iguanodon* was given a nose-horn, like a rhinoceros. A *Megalobatrachus* became a giant, popeyed frog.

The publicity was excellent. Pen sketches of Owen, Hawkins, and twenty distinguished guests banqueting inside the scabrous belly and neck of *Iguanodon* were printed on the front pages of newspapers, magazines, and science journals throughout Europe and North America. British schoolchildren were taken on guided tours to view the "antediluvian monsters" and listen to lectures that craftily evaded a head-on collision with the Scriptures.

Phineas T. Barnum briefly considered commissioning Hawkins to do a second set of dinosaur figures for his American Museum. But the recent concert tour of Jenny Lind had earned him so much profit and publicity that he was thinking about selling the museum and going into politics back home in Bridgeport. Anyway, New York geologists assured him, dinosaurs were a European phenomenon; there wasn't the least shred of evidence that any of these dragons would ever be discovered in the United States.

James Hall was not consulted by Barnum. If he had been, his reply would most likely have been a brusque "No comment." Hall had finally obtained funds from the legislature for a state geological museum. Now he was determined to compare New York's strata and fossil types with those to be found west of the Great Lakes, especially in that vast desert jumble of the Dakotas. There, if anywhere in the United States, there would be evidence of dinosaurs.

And he had the right people for the job. One was the young medical student boarding at his home, Ferdinand Vandiveer Hayden. The other was the young genius teaching anatomy at the University of Pennsylvania, Joseph Leidy.

11 | He Who Picks Up Stones Running

On the 18th we camped near a fine spring. . . . Our animals needed rest, and here was an abundance of good grass and water. After partaking of a delicious dinner of antelope meat, I started out accompanied by my voyageur, and ascending an elevation which was above the bad ground looked down upon one of the grandest views I have ever beheld. The denuded area was nearly square in form, and the immense flat concretions that projected out from the sides of the perpendicular walls in regular seams, and at about equal distances above each other, resembled some vast theatre; indeed, it reminded me of what I had imagined of the amphitheatre of Rome, only nature works upon a far grander scale than man. We climbed with great difficulty down the steep sides, following the main channel of the little stream, and after much winding through this labyrinthian sepulchre, we came to an open plateau covered with fine grass, and in the centre a beautiful grove of cedars, and through the whole a stream of milky water wound its way to White River, about five miles distant. All around us were bare, naked, whitened walls, with now and then a conical pyramid standing alone. We felt very much as though we were in a sepulchre, and indeed we were in a cemetery of pre-Adamite age, for all around us at the base of these walls and pyramids were heads and tails, and fragments of the same, of species which are not known to exist at the present day. We spent that day and the following exploring the cemetery, which the denuding power of water had laid open for our inspection, and many fine specimens rewarded our labors.—Ferdinand V. Hayden's field journal, May, 1855

I F we go up in the Boat to Fort Pierre, we cannot make the Collection and return in three months for less than a thousand dollars," Dr. Ferdinand Hayden wrote, frowning down at the sheet of paper. He stirred the crusted black sludge in the inkpot and stared again out of the inn window to the smokestacks that were suspended, like ebony pipestems, above wisps of fog eddying over the Mississippi River. After a moment, he turned back to the letter he was writing.

"Via boat, the company will furnish our supplies, charge us 50 dollars each for our passage up, charge us for our freight. When we get there, they will furnish us men at from 2 to 3 dollars per day and an old cart that will break down in the first

100 miles, at one dollar per day, 2 mules that have made to flinch out through a cold winter with a small remnant of life left for one dollar per day each. They will expect to furnish food for ourselves and them, at about three times its real value and poor at that. Every thing we get, we must pay at least their prices or they are [un]willing to afford us all the facilities in their power. . . . For anything that they do or furnish, we must pay them an enormous price and make them as little trouble as possible."

It was the night of May 10, 1853. The steamboat stacks, the singing of a roustabout crew, and the fragrances of mud, raw-hide, sweat, charcoal, lilacs, garbage, beer slops, and whore-house patchouli drifting through the curtains of his attic win-dow all attested the Americanization of Saint Louis since Yankee speculators had followed the Lewis and Clark keelboat across the river in 1804.

Keelboats and stern-wheelers, both products of the Philadel-phia waterfront, had been responsible for making Saint Louis the queen of the Mississippi-Missouri watershed; the domain covered more than one-fourth of the land area of the United States and contained a far larger proportion of her natural resources. If and when the railroaders completed the tracks across the Rockies, then Chicago or some other prairie town might divert this tide of wealth and establish its own bedlam. Until that remote time, though, Saint Louis reigned uncon-tested.

The stubbiest stacks on the river were those of the flat-bottomed snag-dodgers that "Old Wienerwurst" Astor had ordered built a generation before for his American Fur Com-pany. The whistles of these boats could, within a month, be echoing at the Yellowstone and Missouri headwaters. Their sole objective was to hasten the plunder of every marketable fur hide out of the Northwest. Any civilian who booked passage on them became merely excess baggage.

The taller stacks identified the lower river fleet. The boats with the slim white ones transported mails, passengers, deli-cacies, and lusty indiscretions between St. Louis, the Ohio valley, Natchez, New Orleans. The big-hipped stacks belonged to the "locals," the lifelines of every plantation and hamlet wharf. Their deck mounds of cotton, groceries, wild honey, beaver pelts, ginseng, sour-mash whiskeys, and farm tools served to hide from view the crap games, singalongs, and unrestrained festivities that characterized life on these boats.

Ferdinand Hayden's concern was access to the western sections of Dakota, where deposits of volcanic ash, fuller's earth, and sand had created a grotesque desert, appropriately called the Badlands. Since his letter focused on the niggardly budgets that would hamper paleontological expeditions in the West until the late 1860's, and inasmuch as it is also a prologue for the most unusual treasure hunts ever undertaken in the West, it is an important American document.

Skinny, somber "Ferd" Hayden was the junior of a two-man expedition that James Hall was sending into the Badlands to study the geology and compare it with that of New York, and to dig out a cartload of the fossils with which, he suspected, its washes and bluffs were pocked.

The decision of a twenty-three-year-old Yankee, just awarded his M.D. by the Albany College of Medicine, to volunteer to assist on this pioneer paleontological venture into the most desolate and dangerous region of the West was a good indication of the excitement that this area of applied science was arousing at the colleges.

Hayden was born on September 7, 1829, on the Connecticut valley's mid-rim at Westfield, Massachusetts. His father died when the boy was ten; his mother remarried and sent him off to live with an uncle who had a farm at Rochester, Ohio. The uncle must have been devoted to "book learning," for, by age sixteen, Hayden had enough of it to be hired as teacher at a nearby school. Though his salary was meager, he managed to save enough to enroll at Oberlin College, twenty-seven miles away. There he earned his board and tuition by moonlighting as a janitor, errand boy, and gardener. He studied so diligently that he was chosen to deliver one of the graduation addresses in

1850. His topic, "The Benefits of a Refined Taste," gave no hint, as a fellow scientist observed decades later, that this "quiet, dreamy, nervous young man would . . . exactly meet the wants of the Great West." Having decided to become a physician, Hayden chose the medical college at Albany, New York, as the one best suited to his needs, and eventually he became a boarder at James Hall's home.

Hall had become as skilled at popularizing fossil collecting as his mentor, Amos Eaton, had been at promoting biological awareness. With appeals based on "New York State prestige" and "educational resources for your grandchildren," Hall had persuaded the merchants, farmers, and attorneys in the legislature to approve appropriations for a state geological museum.

When Hayden showed interest in Hall's personal collection of fossils, Hall sensed another worth-while convert and assembled his lures. Paleontology, Hall maintained, was an exceptionally rewarding field for a medical student, especially one who intended to specialize in surgery. No field of medicine could match the challenges of identifying and assembling the scraps of prehistoric plant life and bone that had been churned, petrified, and tucked away in the depths of the Earth. Caspar Wistar, Gideon Mantell, Richard Owen—the list of physicians who had pioneered in paleontology was impressive. Now, in Philadelphia, the young medical teacher Joseph Leidy showed promise of becoming as great an anatomist of prehistoric life forms as Owen had become in England. There was a real future here, if . . .

The offer of the post of field assistant on the six-month venture into the Badlands was Hall's graduation gift to Hayden in the spring of 1853. Seven years before, a fur trader had brought a fossilized jawbone down the Missouri to Saint Louis and shown it to Dr. Hiram A. Prout. It was, the trapper said, "the chompers" of what the Sioux called the thunder horse, whose ghosts rode from cloud crest to butte to cloud crest on stormy nights. The Makoositcha country between the Missouri's big bend and the South Pass was said to have thousands of bone fossils, "lots of 'em bigger'n this," and mile upon mile of shiny boulders as generously studded with "pewtrified" plants and bugs as a plum pudding is with raisins.

When a copy of *The American Journal of Science* containing Dr. Prout's description of the jawbone fossil and the vast geological dump of the Makoositcha reached Albany, Hall began promoting funds for an expedition there. Some of the rock strata described by Prout seemed to match those in western New York. If this proved true, their plant and animal fossils would be valuable additions to the state museum collections. Joseph Leidy would undertake the diagnosis and might even persuade the Philadelphia Academy to purchase any worthwhile surplus.

In 1849, John Evans had ventured north of the Overland Trail "throughway" up the Platte valley and managed to pacify the Sioux enough to map some of the Badlands and carry out a few more fossils. From Evans and the correspondence of army officers and Missouri valley traders, Hall deduced that the most promising strata were 150 to 200 miles up the White River, almost due north of that massive Overland Trail escarpment called Scotts Bluff. Hall had been able to assemble less than two thousand dollars for the expedition; part of this came from the sale of some of the choicest items in his personal collection of invertebrates and rocks. But to a degree he could make up for that meager sum by assigning his assistant, Fielding B. Meek, to the expedition. An Indiana native, Meek had been an assistant on the first geological surveys of Minnesota so was familiar with both the topography of the northern plains and the arrogance of the Dakotan Sioux.

There could be no salary for Hayden, but, Hall argued, what more profound graduate training for a physician could there be than identifying "pewtrified" chunks of hitherto unknown plants and animals, digging them out of the rock walls, and bringing them safely out of that desolation ruled by Indians long allied to Great Britain?

Hayden was convinced. He left the genteel life of Albany with orders to precede Meek to Saint Louis and to assess the potentials for transportation into the Sioux heartland. He reached Saint Louis a week after the Jesuit missionary Pierre Jean De Smet left for Idaho, and thus missed the opportunity to seek advice from the keenest student of Sioux folkways in all of the West. But he did have the luck to catch up with John Evans, who was outfitting for a trek to Oregon. An evening with Evans, other conferences with the shrewd mountain man and former Indian commissioner, Robert Campbell, convinced him that the Overland Trail and the probable co-operation of army officers would provide Hall with a far greater return than the hard-shell boatmen of the upper Missouri.

"If we go the Land Route," he continued that night of May 10, "we go from here to St. Josephs by steamboat, carrying our provisions and baggage in a good light California wagon. At St. Josephs, we get two good mules, put our effects in our wagon and go with the numerous emigrant trains which constantly fill the road. After leaving St. Josephs our expenses will be nothing, the travel through a fair country in which a thousand things of interest may be found, and as safely as on the Boat, from St. Josephs to 'Badlands' 750 miles, at 25 miles per day, 30 days. . . .

"At Fort Laramie we can get plenty of help cheaper and better than at Fort Pierre. Many of the Soldiers, or even Officers, will like to relieve the tediousness of situation in that way. If not, there are a plenty of [voyageurs] that may be obtained. At Fort Laramie, I am told, we can live much cheaper, and in all probability [it will] cost me nothing to stay through the military. Our 'treasures' we can transport to Fort Pierre and have them sent down by boat or they can be sent from Fort Laramie in the Quartermasters wagons which always return empty. . . . It will cost about 350 dollars to fit out and get started from St. Josephs. After that, the expense will be trifling until we get to Fort Laramie. . . . Dr. Englemann has just enquired the cost of a good wagon. He finds a man who will make a good wagon, fit it all out with cover, etc., warrant it for 70 dollars. He needs but a weeks notice to have it ready. In what way we wish, we need a letter from the Secretary of War to the quartermaster like that one in that little book of instructions of Bairds, and sent on immediately. Dr. Englemann will see that it is sent to Fort Laramie.

"I have written as I think at present. Perhaps when Mr. Meek comes, things will look different."

Either Meek overruled Hayden's choice of the Overland Trail, or Robert Campbell reconsidered and arranged a "cut-rate package." The latter seems probable, because Hayden, Meek, and their gear did buck the Missouri currents and snags to Fort Pierre aboard the *Robert S. Campbell.* They traveled the sixteen hundred miles at an average speed of fifty-five miles a day. Consequently, they were unable to begin the trudge up the valley of the White River until late June.

Bison had hammered a migration trail up the river's red earth trough to summer feeding grounds in the Wind River pastures. Now, a few times each year, supply wagons and horse *remudas* followed this trail between Forts Pierre and Laramie. Antelope, deer, and mountain sheep were plentiful. For a penny or two, Sioux families offered "pick of the garden" from their vegetable patches on the bottomlands. But the only fuels for campfires were dried bison dung and brush. "The soil of these bottoms," Hayden wrote, "is very fertile, composed of the calcerous and aluminous marls of the Tertiary basin through which the upper portion of this river flows, and the clays of the Cretaceous system which forms the hills, and is the basic formation throughout the valley. . . . The water is very peculiar . . . it has much the appearance of milk. When allowed to stand for a short time, or whenever it is found in pools, a thick scum may be seen upon the surface very much of the appearance and consistency of rich cream; removing this, and the thinner portion is of a much lighter color, like milk. It is very astringent to the taste, and its medical effect on the traveller is quite the reverse of the water previously used."

On the tenth day "the tall naked columns and domes of the Badlands" loomed up. "It is only to the geologist," Hayden concluded years later, "that this place can have any permanent attractions. He can wind his way through the . . . canyons among some of the grandest ruins in the world. Indeed, it resembles a gigantic city fallen to decay. Domes, towers, minarets and spires may be seen on every side, which assume a great variety of shapes when viewed in the distance. Not unfrequently, the rising or the setting sun will light up these grand old ruins with a wild, strange beauty, reminding one of a city illuminated in the night when seen from some high point. . . . It is at the foot of these apparent architectural ruins that the curious fossil treasures are found."

They were, the travelers slowly deduced, on the site of a titanic battle. Millions of years before, when this bleakness was rich pasture land bordering a lake, strange families of carnivores and ruminants here waged a fierce struggle for survival. Many of the ruminants resembled pigs, although some had been as large as ponies. They ran in vast herds, very much as the bison now did. Their enemies were saber-toothed tigers, wolves, and immense hyenalike creatures. Mingled with their skeletons were the shells of countless turtles, varying in size "from an inch or two in length across the back to three or four feet." It would have been impossible to believe, if the rock strata had not provided the evidence, that the bones had been there for millions of years, since the middle Tertiary period. "The bones are so clean and white and the teeth so perfect," Hayden wrote in his journal, "that, when exposed upon the surface, they present the appearance of having bleached only for a season."

The problem was not searching but selecting, and Meek had more experience in that than Hayden. So, while Meek pondered, sketched, and wrote descriptions of the strata and their treasures, Hayden explored the canyon floors and walls. He began to discern orderly patterns of time and natural catastrophes from the medley of spires, buttes, and gorges. By and large, he decided, there were three types of rock formation. Each had its own graveyards of distinctive bones and teeth; many of them were species he had never seen in the Albany collections or heard about from Dr. Hall. He could not, he finally decided, abandon this amphitheater of the Earth's ancient struggles after the few weeks he and Meek would be able to spend.

The details of the young physician's struggles with his conscience and with Hall and Meek during the next eighteen months were not adequately recorded. By the time Hayden and Meek returned to Albany, Colonel Ezekial Jewett, still interested in fossils after a vain search for gold in California, was bargaining with Hall to become curator of the state museum. Perhaps Jewett, as well as Hall, persuaded Hayden to seek funds from Joseph Leidy. Letters from Leidy indicate that during the winter of 1853–54 Hayden sent Leidy some of the choicest bones he had found in the Badlands and began the correspondence that would play a critical role not only in both of their careers but in the future of paleontology in America.

Confident that sales to Leidy and other fossil collectors would provide the few dollars necessary to grubstake him, Hayden returned to Saint Louis in the spring of 1854, shipped up the Missouri as a deck hand, and jumped ship at or near Fort Pierre. He could not even afford a mule and chose not to carry a gun. Supplied only with a knapsack, a pick, and a specimen bag, he hiked west to the Badlands and Black Hills. There, sunburned to the copper hue of a Sioux, black beard ruffled by the gully winds, he explored each gorge to puzzle out the pattern of the strata. The Sioux farmers and foragers he encountered, recognizing that he presented no threat to them, fed him, gave him bits of fossils, and directed him to deposits of the "ghost horses." They were so impressed by his intensity that they nicknamed him "He Who Picks Up Stones Running."

It must have been a Sioux, or a half-breed trapper, who first told Hayden about the "pewtrified trees and critter bones" protruding from the ledges along the upper Missouri and Judith rivers, far north of the Black Hills, almost at the Missouri's Great Falls in Montana. The Judith and its spectacular basin was in hunting lands claimed by Sioux, Crow, and Blackfoot, and was remote from any protection by army patrols out of Fort Benton or Fort Belknap. Sometime during the fall of 1854, Hayden reached the Judith country, recognized its outcroppings as Mesozoic, and found the first dinosaur teeth ever identified as "native North American."

Back in Saint Louis by late fall, he wrote Joseph Leidy, offering him the best "fossil remains" from his Badlands–Judith River collections. He then began a series of interviews with politicians, army officers, and traders, in which he suggested a variety of schemes, including mineral surveys, to finance a more elaborate expedition into the Badlands in 1855. He found a benefactor in Colonel A. J. Vaughan, an Indian agent and a collector of geological specimens. Colonel Vaughan's cash advance against first choice of the year's finds paid for a string of burros, some surveying instruments, and Indian hunter-guides.

A few weeks after the party left Saint Louis, Leidy wrote an eager offer to "examine and describe the new vertebrate remains you have discovered," and to "purchase a set for the Academy from which I would try and obtain the highest price on the value. . . . Lastly, as a tribute of respect to your zeal I have proposed you as a correspondent to our Academy and hope the next time of writing to announce your election."

The summer-long search into the Badlands via the valleys of the Bad and White rivers, with a second foray into the Judith region, yielded a dozen crates of bones and teeth. After Colonel Vaughan chose his favorites, Hayden sold a collection to the new Saint Louis Academy of Sciences. But the choicest fragments from the upper Missouri and Judith beds were shipped to Leidy. That winter, Hayden rode the new railroad trains to Washington, where, thanks to Leidy's introductions, Joseph Henry and others at the Smithsonian Institution welcomed him and initiated the "politicking" to obtain a position for Hayden on a topographical survey of the Black Hills to be undertaken by and for the War Department in 1856.

On February 25, 1856, Leidy wrote excitedly that the box "containing saurian remains [included] a parcel of fragments of dense rib-like bones with several fragmentary bodies of vertebrae. These I suspect to have belonged to a herbivorous cetacean of remarkable character. In what formation were they found? The vertebrae and other bones embedded in two large masses of hard matrix is a new and very peculiar genus of ichthyoid reptilia. . . . Where is the Judith River of which you speak? A branch of what larger stream is it?"

The years 1856, 1857, and 1858 were eventful ones for the "Great American Desert." Proslavers and abolitionists battled for domination of the Kansas and Nebraska territories; the massacre of five proslavers on Pottawatomie Creek gave John Brown national notoriety. The completion of railroad links between New York, Chicago, and Saint Louis encouraged "Crazy Ted" Judah to sail from San Francisco to Washington in order to distribute his tiny pamphlet, "A Practical Plan for Building the Pacific Rail-road," to congressmen. The discovery of gold in Colorado and Minnesota's admission to the Union each instigated a "rush" of settlers and promoters. President Buchanan sent an expeditionary force into Utah to "chastise the ungodly, polygamous Mormons"; James Butler Hickok and William Frederick Cody were among the youthful drivers of one of the expeditionary force's supply trains. The burning of their train by Mormon raiders had a direct bearing on the decision to inaugurate the Pony Express three years later.

But none of these events was more important to America's maturing than the fossil hunts and geological surveys of He Who Picks up Stones Running. As official geologist with Lieutenant G. K. Warren's topographical survey of the upper Missouri and the Black Hills in 1856–57, Hayden returned to his beloved Judith River wonderland. Later he followed the Platte's prairie swirl across Nebraska to Fort Laramie. For the first time, he examined the tawny upheaval of "basement rock" in South Pass, the Wind River cliffs gleaming with petrified fish scales and blue-flecked groves of agatized trees, and the evidences of mineral wealth in the Black Hills. "He worked out the broad geologic outlines of the Black Hills," Gilbert F. Stucker wrote more than a century later, "determined their domal structure, discovered the first Jurassic beds on the continent on their flanks, and collected fossil evidence to prove the existence of Cambrian strata—those then oldest-known fossiliferous rocks—west of the Missouri River."

The bundles Hayden shipped to Philadelphia continued to elicit praise from Leidy. "After some conversation with members of the Academy," he wrote in January, 1857, "I am at liberty to offer you for it, the money intended towards an expedition, the sum of $1200. . . . In regard to publishing, I think you had better furnish me with some of your geological notes, and permit me to prepare a memoir for the Smithsonian immediately."

Four months later, Leidy warned Hayden that "the British Government has formed an exploring party to determine a route to Vancouver's Island. They are to go west from the headwaters of Lake Superior, and thus may walk right into the affections of your Miss Judith. What is to be done?"

"I have commenced examining the fossils last sent and, as you anticipated, I have been surprised," Leidy wrote in January, 1858. "The fauna is so different from that of the Mauvaises Terres. . . . The collection contains fragments of about twenty different animals, all I think different from those of the Nebraska Miocene. . . . This new Pliocene collection only makes me thirst for more, as there are many fragments of carnivora and pachyderms quite evident but not sufficiently characteristic for description. I hope you have it in your heart again to visit this locality as well as the one of Judith River."

With Leidy as editor, the Philadelphia Academy published *Descriptions of New Fossil Species of Mollusca Collected by Dr. F. V. Hayden in Nebraska Territory* in November, 1856, followed by the Hayden-Meek *Notes on the Geology of the Mauvaises Terres of the White River, Nebraska* and *Explorations Under the War Department* in the spring of 1857. These papers brought Hayden invitations to lecture before the American Philosophical Society, to write for "Silliman's Journal," and, most important, strengthened Hayden's relationships with Joseph Henry and other executives at the Smithsonian Institution.

Proud of his protégé, James Hall gave Fielding Meek the

summer off in 1858 so that Meek could join Hayden in another
search, this time into western Kansas for the General Land
Office. Hayden's geological timetable of the rock strata in the
Badlands had been published in his 1857 report for the Warren
survey; it would prove to be the basic source for paleontological and geological data on the region. Now he was intent on
similar trail blazing through the mysteries of the Wind River
range and those sepulchral cliffs that provide the rivulets of the
Yellowstone's headwaters.

The opportunity came that winter, when Leidy and the
Smithsonian's officials succeeded in having Hayden appointed
geologist of the War Department's topographical expedition to
the Yellowstone headwaters. Captain W. F. Raynolds commanded the pack-pony brigade that shuffled north from Fort
Laramie in the late spring of 1859.

The guide was fifty-five-year-old Jim Bridger, already a
legendary figure. A farm boy from the fringes of what is now
Kansas City, Jim had been the youngest member of Jedediah
Smith's 1823–24 expedition that discovered the Sweetwater
River approach to the Rockies' most important crossing place,
South Pass. Mountain man, adviser and guide to Brigham
Young during the first Mormon migrations, Bridger was also
credited with originating the stories about "pewtrified birds in
pewtrified trees."

Hayden would one day discover that there was little exaggera-
tion in those stories; Jim had merely reported what he and
other mountain men had gawked at amidst the boiling springs,
geysers, and giant stalactites around the falls of the Yellowstone. Hayden and his associates would value Jim's know-how
enough to memorialize him by coining the term "Bridger
green" to describe the jade-green outcroppings along the Sandy
and Green rivers.

The mountain passes that led to the geological phenomena
Jim Bridger had hoped to show Hayden in the geyser basin
during 1859–60 were blocked by heavy snows and could not
be negotiated. A decade and more, plus a war, were to elapse
before Hayden saw them, and, when he did, they would have

Dr. Caspar Wistar

Charles Willson Peale, self-portrait

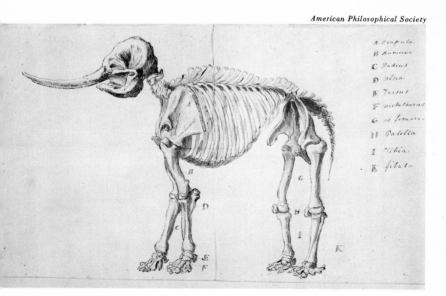

The Peale "mammoth": one of two skeletons constructed by the
Peales and Wistar after the 1802 Wallkill River swamp digs

Exhuming the First American Mastodon by Charles Willson Peale, 1806–08

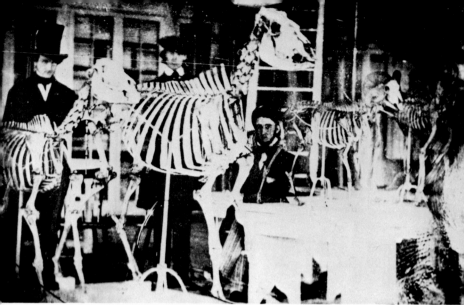

Fossil skeletons on display at the museum library, Philadelphia
Academy of Natural Sciences, about 1840.
Edgar Allan Poe (?) at left

Amos Eaton   James Hall

Erie Canal construction, 1820–24, at Lockport, through ridge diking Lake Erie above Lake Ontario. The ridge's dolomite and shale are studded with fossil shells and bones. Eaton, Lyell, and Hall collected here, as did Ezekial Jewett, who gave Othniel Marsh his first paleontological lessons at this site.

The Erie Canal as Eaton, Hall, Lyell, Powell, and Marsh knew it. *A Sultry Calm. Pitsford* [sic] *on the Erie Canal* by George Harvey

Joseph Leidy

Dr. Ferdinand V. Hayden

Edward Drinker Cope

Othniel C. Marsh

The 1870 Hayden expedition. Hayden is seated at the far end of the table, facing camera. Photographer William H. Jackson is at the far right.

Major John Wesley Powell, the ablest director of the Geological Survey, talking to a Paiute Indian about the location of water during the northern Arizona survey, 1873

Dr. Jacob Wortman (left) and O. A. Peterson, a paleontological
prospecting outfit, Kinneys Ranch, Wyoming, 1890's

Elmer Riggs and Barnum Brown at the *Coryphodon* beds,
Wasatch badlands, near Otto, Wyoming, 1890's

Backbone of a huge *Ichthyosaur* uncovered in the Rattle Snake Hills, near the Oregon Trail route, central Wyoming, 1890's

Dinosaur bones in position, Bone Cabin Quarry, Wyoming, 1898

Othniel Marsh's greatest memorial, Dinosaur Hall at the Peabody Museum, Yale University

Dinosaur bone fossils at Como Bluff, 1898–1900

Hauling fossil bones to the railroad from camp at the mouth of Sand Creek on the Red Deer River, Alberta, Canada, about 1912

W. H. "Bill" Reed

Henry Fairfield Osborn

Dr. George Gaylord Simpson

Dr. Walter Granger

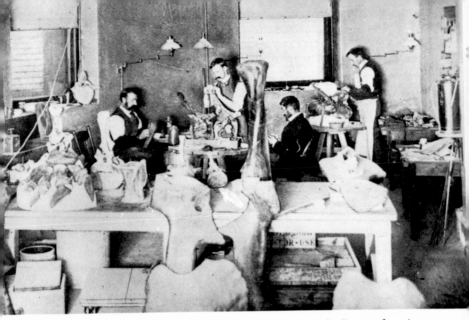

Dinosaur laboratory, Carnegie Museum, Pittsburgh, Pennsylvania, 1899. Left to right: W. H. "Bill" Reed, Arthur Coggeshall, Jacob Wortman, O. A. Peterson

Carl Sorensen working on *Coryphodon* block, 1950's

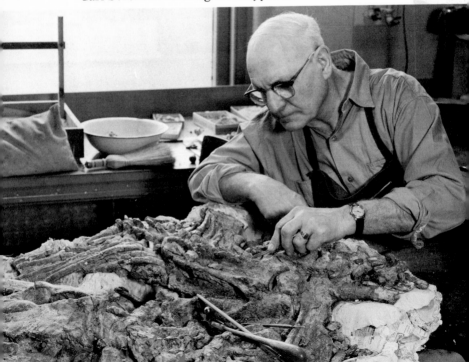

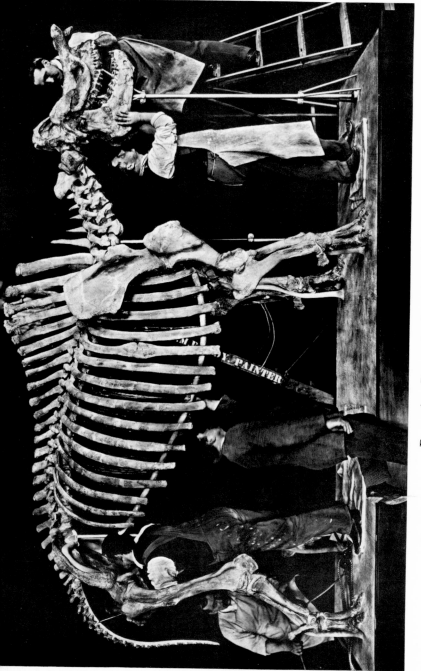

Preparing *Brontotherium gigas*, 1900

The Dinosaur National Monument when the Carnegie Museum (Pittsburgh) was working there. Vernal, Utah, about 1910

Original Dinosaur Quarry cut. Looking east, showing part of quarry excavated by the Carnegie and National (Smithsonian) museums. Dinosaur National Monument, 1910–20

Earl Douglass, discoverer
of the bone quarry at the
Dinosaur National
Monument, Vernal, Utah

National Park Service paleontologists—Gilbert Stucker (top left),
"Tobe" Williams (bottom left), George Robinson (right)—working at
the quarry face, Dinosaur National Monument, mid-1950's, before its
enclosure in the exhibition building

U.S. Department of the Interior, National Park Service

Quarry technician Jim Adams finishing exposure of a mass of fossil
bones from the dinosaur *Camarasaurus supremus*,
Dinosaur National Monument, 1970's

Spectator gallery at quarry face, Dinosaur National Monument, 1970

U.S. Department of the Interior, National Park Service

impact on the social policies of the federal and state governments. The spectacle of the primordial forces still at work along the Yellowstone headwaters would convince Hayden that lobbying should be initiated to preserve them, "in the public interest."

Meanwhile, the fossils and reports shipped back to Washington and Philadelphia substantiated the suspicions of Joseph Leidy and the Smithsonian scientists that the wilderness west of the Missouri contained the most significant array of prehistoric fossils ever discovered.

By the time the Raynolds expedition returned to Fort Laramie, the Pony Express was clattering between Saint Joseph and Sacramento, and there were rumors of secessionist plots in California, Nevada, and Colorado. But the packets of scientific journals waiting for Hayden teemed with articles, denunciations, and defenses of a matter as significant to geologists as the possibility of Abraham Lincoln's election was to other Americans. Charles Lyell's friend and confidante, Charles Darwin, had been persuaded to publish *The Origin of Species*. And now the battle lines were forming. There might be a War Between the States in America; but there was certain to be another war—between science and religion—everywhere else.

12 | Darwin's War

We contemplate with satisfaction the law by which in our long history one religion has driven out another, as one hypothesis supplants another in astronomy or mathematics. The faith that needs the fewest altars, the hypothesis that leaves least unexplained, survives; and the intelligence that changes most fears into opportunity is most divine. We believe this beneficent operation of intelligence was swerving not one degree from its ancient course when under the name of scientific spirit, it once more laid its influence upon religion. If the shock here seemed too violent, if the purpose of intelligence here seemed to be not revision but contradiction, it was only because religion was invited to digest an unusually large amount of intelligence all at once. Moreover, it is not certain that devout people were more shocked by Darwinism than the pious mariners were by the first boat that could tack . . . if intelligence begins in a pang, it proceeds to a vision.—John Erskine, "The Moral Obligation to be Intelligent," 1912

BETWEEN 1861 and 1865, the Potomac River became the blood-stained frontier of the torn American Union. Urged on by the assurances of chaplains (both fundamentalist and radical) and generals (both professional and political) along both shores that "God is with us," more than one million men deployed, fired, bayoneted through a series of battles, some more bloody than the world had ever known, at Harpers Ferry, Manassas, Antietam, Winchester, Gettysburg, and other communities on the river's watershed.

In February, 1863, eleven Protestant denominations sent delegates to a conference at Xenia, Ohio, to formulate a petition asking those States still faithful to the Union to start the ponderous machinery that would affix an amendment to the Constitution professing "allegiance to God." From this came the National Reform Association, financed largely by Presbyterians and Episcopalians, which was dedicated to agitating and lobbying for such a "Christian amendment," with sublime indifference toward the First Amendment's requirement that "Congress shall make no law respecting an establishment of religion, or prohibiting the free exercise thereof."

Extraordinary skill in implementing mass murders, looting, and tortures in the name of "meekandblessedJesusamen" had shaped the raw and ugly history of Europe for more than eighteen hundred years. It was logical that the "staunch" Presbyterians, Methodists, Baptists, and Episcopalians who were bickering for leadership of the Union and Confederate armies had no difficulty in assuming, through their chaplains and bishops, the divine right to communicate with Almighty God at will and suspend the pacifist teachings of Jesus for the duration of the fratricidal war. From Constantine the Great through Joan of Arc to Napoleon and Blücher and Wellington this pattern of nationally adjustable piety had been the mode.

However, on both sides of the Potomac, the spasm of self-serving piety became more intense than it had been during the Revolution, the War of 1812, or the expansionist Mexican War. Not all of this flowed from the Sunday-morning meekness of the generals, politicos, and profiteers. A second and more durable war was developing, triggered by a book written by Thomas Lyell's old friend, Charles Darwin.

During 1857 and 1858, Darwin swiftly organized the statistics and insights accumulated during his thirty years of analysis of geological, botanical, and zoological evidence. With surgeon-like skill, he excised folklore and mythology and exposed the intricate, immeasurably ancient pattern of the evolution of the Earth's life forms from one-celled globs to the contemporary array of plants, animals, and fungi. Mankind, he assumed, was also a product of this vast transmutation. It seemed probable that at some remote time, *Homo sapiens* temporarily looked and thought and behaved like an ape.

Darwin titled his work *The Origin of Species*. It was published in London in 1858, during the months when the Lincoln-Douglas debates at Illinois county seats were providing the lank, homely Springfield attorney with his first nationwide publicity.

Ever since the arguments about the *Mammut* skeleton and the 1795 publication of James Hutton's *Theory of the Earth*, clergymen and philosophers had been forced to retreat, regroup, retreat from the medieval notion of a finite God who, during one whimsical week in 4004 B.C., created the planet Earth and fashioned man and woman out of two sterile mud balls. William Smith's stratigraphy, Baron Cuvier's catastrophism, James Lyell's *Principles of Geology*, Richard Owen's identification of the huge "prehistoric lizards" had each been a disastrous defeat for the advocates of recent Creation. And now the reclusive Darwin seemed to be tearing at the very keystone of Christianity's arch by suggesting that, inasmuch as it was inconceivable for Almighty God to look like a polliwog or an ape, *Homo sapiens* had *not* been created "in God's Image."

The first London printing of *The Origin of Species* was modest. Author and publisher agreed that there would be no more than a slow, hopefully steady sale to botanists, zoologists, paleontologists, perhaps a few museum libraries in Great Britain and America. Thomas Lyell read the galley proofs and congratulated Darwin, but he decided that he could not accept the theory of evolution, especially that ape phase. Dean Buckland, having blithely relocated the Garden of Eden to a fairly recent afterthought by the Almighty, was forced into reluctant agreement with the most ardent defenders of Bishop Ussher. Vituperative reviews, anguished sermons of denunciation, and the necessity for every scientist, clergyman, curator, teacher to take a firm pro or con stand all combined to turn the book into a runaway best seller and establish it as "the most important publication of the century."

A climactic confrontation occurred during the June, 1860, convention at Oxford of the British Association for the Advancement of Science. Three years before, Richard Owen had been promoted to the superintendency of the natural history division of the British Museum and had been knighted by Queen Victoria. Sir Richard was very conscious of his obligations to the Church of England as well as to the museum when he climbed the podium steps on the first afternoon of the BAAS convention. Nineteen years before, he had made his historic pronouncement about the dinosaurs at Plymouth. At that time, he had confused and angered many church dignitaries. Now his course was clear. He would stand with the church. "The brain of the gorilla," he pronounced, "presents more differences, as compared with the brain of man, than it does when compared with the brains of the very lowest and most problematical of the quadrumana."

Soon after making this statement, Sir Richard retired to a conference room with "Soapy Sam" Wilberforce, the bishop of Oxford, to serve as one of the ghost writers of a scathing indictment of Charles Darwin and his "heretic theories" that the bishop planned to deliver before the convention.

Darwin was either too shy or too ill to attend the meeting. But he found a champion in Thomas H. Huxley, a waspish surgeon turned writer and lecturer. Huxley's cogency in a rebuttal to Wilberforce's attack on "mankind's descent from the monkey" evoked both catcalls and cheers from the seven hundred scientists and clergymen crowded into the lecture room. Subsequent debates, lectures, and essays by Huxley and his close friend, Herbert Spencer, established them as the most persuasive disciples of Darwinism. Huxley coined the word "agnostic" (from the Greek *agnostos,* meaning "not known") to identify people who, like himself, believed that "the Ultimate Cause and the essential nature of things are unknown and unknowable."

This deliberate intellectual modesty, so drastically opposed to the "blind faith" that had been demanded by Christian clergy for eighteen hundred years, received so much publicity that, in 1864, Benjamin Disraeli deemed it politically expedient to announce his convictions. He provided the anti-Darwinists with one of their choicest clichés when, during a speech at Oxford, he declaimed, "What is the question now placed before society with a glib assurance the most astounding? The question is this: Is man an ape or an angel? Now I am on the side of the angels."

Neither Darwin nor Huxley nor Spencer denied the existence of God, although Huxley preferred the term "Ultimate Cause" and Mrs. Darwin despaired that "God is being moved farther and farther from us." However, the historic interpretations of

Genesis and the assumption that man had been conceived "in
God's image" were so deeply entrenched in the Judeo-Christian
psyche that charges of "infidel" and "atheist" were soon lodged
against Darwin and his followers. Two basic positions emerged:
the anti-Darwinists, who insisted on a finite God who had fairly
recently, and quite whimsically, created the planet Earth and
the in-His-image human being during a 144-hour period; and
the Darwinists, who contended that God or the Ultimate Cause
was infinite.

In the United States, President Lincoln deemed it prudent to
incorporate occasional references to "the Almighty" in
speeches and proclamations. From the sweatshops of Manhattan
to the brothels of San Francisco, strident protestations of faith
in the Almighty arose, punctuated with cries of "In Your
Image" and "No apes in the family tree."

The enemies of Darwin's theory took the collective name of
fundamentalists. On the issue of "God image" versus "ape
phase," Catholics and Protestants would labor together more
harmoniously than they ever had. Ironically, before the Civil
War ended, the Puritan stronghold of Harvard University became campaign headquarters for both the fundamentalists and
the Darwinists. Professor Asa Gray was the American champion of Darwinism; Professor Louis Agassiz was the most
astute supporter of fundamentalism.

Amos Eaton's friendship with the eager farm boy from
Oneida had had momentous effect. Through Eaton, Asa Gray
met John Torrey, became his assistant, searched the botanical
mysteries of the trans-Mississippi West with him, and by 1842
was professor of natural history at Harvard. His *Manual of
Botany*, published in 1850, became, and still is, the standard
reference work on the flora between the Atlantic and the
Rockies.

Charles Darwin's search for data about the variations and
natural-selection processes of Earth's life forms brought him
into correspondence with Gray during the early 1850's. Gray
became Darwin's most intimate correspondent in the United
States and the foremost American authority on the intricacies of

and biological evidence for the theory of evolution. His magnificent review of *The Origin of Species* in "Silliman's
Journal" won James Dwight Dana, Silliman's son-in-law and
the editor of the journal, over to the side of the Darwinists.
Through it all, Professor Gray remained a church member and
a devout believer in the infinity and majesty of God.

"I am bound to stick up for [Darwin's] philosophy," Gray
explained to the physician-botanist Francis Boott on January
16, 1860. "I am struck with the great ability of [his] book and
charmed with its fairness. I also wanted to stop Agassiz's mouth
with his own words, and to show up his loose way of putting
things. He is a sort of demagogue, and always talks to the
rabble."

Dr. Gray's petulance was not surprising. The enmity between
the two Harvard professors had been growing for more than a
decade. In 1850, two years after Harvard lured him away from
James Hall's staff at Albany, Agassiz married Elizabeth Cabot
Cary, a literate Back Bay Brahmin who was a staunch supporter of women's rights. Her speeches and drawing-room politicking for a "female institution" at Harvard would make her,
in 1879, the logical choice as Radcliffe's first president.
Agassiz's son, Alexander, came to Harvard from Switzerland in
1849, earned a B.S. degree under Gray, then collaborated with
his new stepmother in researching and writing *Seaside Studies
in Natural History*. Meanwhile, his father had adopted James
Hall's technique of trading for fossils and rare rock specimens;
the hobby led to the creation of Harvard's Museum of Comparative Zoology.

The intellectual and entrepreneurial brilliance of the Agas-
sizes, and Mrs. Agassiz's Cabot heritage, made the family as
welcome at political rallies and athenaeum lectures as they
were at the soirees and missionary benefits of Back Bay and
Chestnut Hill society. The publicity they received irked Gray
and may very well have prompted him to draw Agassiz into the
feud over Darwinism and "the monkey business."

Agassiz professed that new species could originate only
through the direct intervention of the Almighty. "If we can

prove premeditation prior to the act of creation," he wrote in the introduction to his *Contributions to the Natural History of the United States of America* (1857), "we have done once and forever with the desolate theory which refers to the laws of matter as accounting for all the wonders of the universe, and leaves us with no God."

Gray's review of *Origin of Species* was published in "Silliman's Journal" in February, 1860; Darwin pronounced it "admirable; by far the best which I have read." A few issues later, Dana gave Agassiz equal space for his review of the book. Agassiz cited paleontology, geology, zoology to demonstrate that "Darwin's theory is thus far merely conjectural." Then, just as "Soapy Sam" Wilberforce had recruited Sir Richard Owen, Horace Greeley asked Agassiz to write a series of articles "demolishing the Darwin theory" in his *New York Tribune*. Agassiz accepted, and Greeley plugged the articles through bombastic advertising in *The Nation*. Notoriously weak about resisting opportunities for publicity, Oliver Wendell Holmes, Sr., soon publicly expressed his thanks that "so profound a student of nature as Mr. Agassiz has tracked the warm footprints of divinity throughout all the vestiges of creation."

By the end of the Civil War, religious magazines were loudly proclaiming that "belief in Evolution is a denial of God." The reactions ranged from avowals of atheism by such "radicals" as Robert G. Ingersoll, Anna Eastman, and Yale's William Graham Sumner, to the stern fundamentalist insistence on a six-thousand-year-old Earth by Seventh-Day Adventists, Mormons, Southern Baptists, and sundry new evangelical sects.

The dilemma of "God versus Chaos," as some scholars would label Darwin's war, was that science had become so critical a force in the American economy that neither industry nor government nor the churches could afford to abandon it or deny most of its techniques. Between 1790 and 1867, application of scientific principles led to the invention of the steam engine, the steamboat, the steam train, the reaper, the air drill, the telegraph, the elevator, and the steel mill, as well as an amazing array of such essential gadgetry as barbed wire, nail mills, horseshoe mills, windmills, McAdam's Asphalt Compound, the kerosene lamp, and the coal stove. Each of these inventions increased the need for raw materials, hence made the geologist, chemist, physicist, botanist, and zoologist essential adjuncts of industry. Yet the "laws of nature" these scientists were using to locate supplies of minerals and oil, to increase crop and livestock yields, and to ascertain the efficiency of new machines and processes were the very laws that Darwin had used in arriving at his theory of evolution.

The increasing dependence of industry on science became obvious to even the most ardent fundamentalists when, a few weeks after Lee's surrender and Lincoln's assassination, construction of the first transcontinental railroads finally began to hammer "high iron" across Nebraska and the Sierras. Geological surveys were essential, for the railroad-construction companies had to find out whether there were potential sources of oil, coal, iron, copper, and other minerals along the rights of way. Next came teams of botanists, zoologists, and chemists to determine the agricultural potentials of the millions of acres being awarded to the construction companies by the federal and state governments. In short, science was about to transform that midriff of the United States long labeled "the Great American Desert." There would be vast spoils for the participants. Neither the pastor of Midcity Temple nor his banker-engineer-manufacturer deacons wished to strangle the goose, so avid were they for the golden eggs.

Thus theologians once again revised their convictions. Father A. F. Hewitt, one of the founders of the Paulists, stated the new position most succinctly in 1887: "The hypothesis of evolution must stop short of Man." This would, in effect, become the official position of the Roman Catholic church, and of several large Protestant denominations. Still, as recently as 1972, fundamentalists in California could exert enough political muscle to require that evolution be taught only as "a theory" in state-supported schools and universities.

By 1866, the young science of paleontology was again at a crossroads. This study of fossils had, since 1800, substantiated geology as an exact science, had sired the museum and educational studies and research in natural history, and had shepherded the transition from "philosopher" to "scientist." Now it provided the soundest techniques for proving or disproving the theory of evolution by substantiating data about the eras of the Earth and concurrent changes in life forms. This heralded two opportunities and produced a new offspring in the sciences.

Research about the dawn history of *Homo sapiens* had to be professionally segregated from research about other fossils. Expediency and human self-respect both demanded it. In 1857, German paleontologists had unearthed a strangely shaped human skull in the Neander valley. A variety of guesses about Neanderthal man and his cave life was appearing in both popular and scientific publications. Searches were under way for other "humanoid" links between Darwin's ape and *Homo sapiens*. Thus one of the ribs of paleontology became the new science of *anthropology*.

But the operation was painless. The transcontinental railroads now offered speedy transportation to and freighting from those vast fossil beds that Ferdinand Hayden and a few others had opened on the Rocky Mountain slopes. And in New York City, a young naturalist from Maine was demonstrating how to lure funds for larger and better museums away from millionaires eager for cultural status.

13 | The Innovators

A number of gentlemen have long desired that a great Museum of Natural History should be established in Central Park, and having now the opportunity of securing a rare and very valuable collection as a nucleus of such Museum, the undersigned wish to enquire if you are disposed to provide for its reception and development.
—Petition to the commissioners of Central Park, New York City, signed by Theodore Roosevelt, Sr., William E. Dodge, Robert Colgate, Levi P. Morton, J. P. Morgan, Alexander T. Stewart, and others, December, 1868

THE "glory of the Coming of the Lord" envisaged by Julia Ward Howe produced, between 1861 and 1865, 500,000 war dead, ten ravaged states in the Southeast, bandit gangs of war-shocked veterans in the West, a surge of migration to northern cities, and scores of *nouveaux riches* whose wives and children were hungry for social status.

While the South was taken over by the carpetbaggers and Ku Klux Klansmen of Reconstruction, while the West felt the first shock waves of technology and industrial exploitation, and while the North underwent social, economic, and political centralization, the new millionaires learned that social status was available through the simple act of donating money to museums, art galleries, and public libraries. Gifts of bonds and stocks, presentations of collections of gems, butterflies, paintings, and curios would be considered "benefactions for the public good" and lead to such ego-serving rewards as honorary degrees, memorial buildings, cornerstone plaques, and brass busts. Awareness of this simple technique spread rapidly between 1870 and 1890 and, after 1900, spawned such powerful institutions as the trust fund and the foundation.

For the museums and other institutional recipients, the rewards of this new trend went beyond financial gain. The donors applied their organizational skills, insisted on "business methods," placated bickering scientists, and strenuously championed the right of the "unwashed masses" to examine and comprehend the wonders of natural history. A few millionaires, notably New York's Morris K. Jesup, became so involved with the museums they were supporting that they took over much of the administration and personally arbitrated everything from salaries for janitors and locations of toilets to begging free passes from the railroads for field expeditions. The second and third generations of benefactory families produced some of the most notable scientists and museum developers of the twentieth century.

The impact on paleontology was momentous. There were now sufficient funds, administrative skills, and political support for innovations made timely by Darwinism, the new public appetite for "exhibitions" of every kind, and the promise of fossil treasure troves in the trans-Missouri West.

By 1865, the remnants of the Peale Museum were scattered in basements, theater lobbies, and blowsy side shows from South Carolina to Boston. The battered skeleton of the great mastodon occupied a dark corner of the museum in Darmstadt, Hesse, only 150 miles southwest of Christian Michaelis's home town. The dioramas of natural habitats and the lifelike taxidermy so vigorously insisted on by Peale had been generally abandoned, not without reason. The fakery and sensationalism used by P. T. Barnum and his medicine-show imitators with their "mummied mermaid under glass bell" and "two-headed man" exhibits caused curators and scientists at such institutions as the Smithsonian and the Philadelphia Academy to react against "excessive popularization" and "pandering to the rabble." Thus the exhibit of a dinosaur leg finally assembled by Joseph Leidy in Philadelphia was a stark bones-and-plaster presentation.

Similarly austere settings were the rule at the hundreds of curio repositories at universities and county-seat exhibition rooms. Budgets were too skimpy for the research and artistry necessary to create the realistic displays Peale had advocated. Nor, so far as paleontologic materials were concerned, was there enough knowledge of prehistoric environments to avoid the grotesque errors Hawkins and Owen had made in their Crystal Palace statuary.

In 1875, Colonel Jewett's protégé, Othniel Marsh, summarized the challenges of both environmental and technological research. Endowments by his millionaire uncle, George Peabody, had expedited his appointment as Yale's first professor of paleontology; fossil hunts into Hayden's "pre-Adamite cemeteries" were establishing him as an authority. So the secretary of the Smithsonian Institution asked him about the advisability of "a display of prehistoric animals" at the 1876 Centennial Exposition in Philadelphia.

"I do not believe it possible at present," Marsh replied, "to make restorations of any of the more important extinct animals of this country that will be of real value to science, or to the public. In a few cases where the material exists for the restoration of a skeleton alone, these materials have not yet been worked out with sufficient care to make such a restoration perfectly satisfactory. . . . Where the skeleton is only partially known, the danger of error is of course much greater, and I think it would be very unwise to attempt restoration as error in a case of this kind is very difficult to eradicate from the public mind. . . . A few years hence we shall certainly have the material for some good restorations of our wonderful extinct animals, but the time is not yet."

This opinion was not only a forecast of the revolutionary techniques that paleontology would impose on the world's museums between 1880 and 1930, but was indicative of the individual dedication that would search out dawn-age artifacts and solve the problems of transferring them to accurate dioramas, paintings, sculpture, and other media.

The vigorous opposition of fundamentalists in Congress and the state governments would so impede the flow of tax money to paleontological research during these fifty years that the skills of prospecting, extracting, assembling, and displaying "the wonderful extinct animals" and their environments would be supported mainly by millionaire patrons and hobbyists, notably Andrew Carnegie, J. P. Morgan, Alexander Stewart, Morris K.

Jesup, Marshall Field, and Othniel Marsh's uncle, George F. Peabody.

One of the most important pioneers of museum innovations was a shipbuilder's son named Albert S. Bickmore. Born in Tenant's Harbor, Maine, Bickmore was sent off to Dartmouth College in 1856 and there became intrigued by mineralogy and geology. In the fall of 1860, he enrolled at Harvard for Louis Agassiz's graduate courses in natural history.

Agassiz was at that time engrossed in the battle against Darwinism. Also, he believed—as many professors still believe —that researching and ghosting first drafts of articles and books he intended to publish was "excellent experience" for graduate students. Bickmore displayed such skill and enthusiasm that Agassiz appointed the twenty-one-year-old as one of his assistants "to help take care of the Radiates and Mollusks" at the Museum of Comparative Zoology.

Peace and harmony apparently prevailed until the summer of 1863, when Agassiz denied Bickmore and other assistants permission to publish research papers under their own names. The following December, Agassiz refused to renew Bickmore's appointment. Bickmore left Harvard at once, after bragging that he would found a museum of natural history in New York City that "will be run on democratic principles."

But Manhattan was still uncertain territory for a young man in "civvies," particularly in a greatcoat and top hat. Six months earlier, with the effectiveness of the Union victory at Gettysburg still in question, the draft riots had erupted in New York City. More than a thousand men, women, and children—mostly blacks—were lynched, shot, or stoned to death within a week. The principal grievance of the rioters was the government's policy of accepting cash payments in lieu of military service, and the resultant scarcity of "rich brats" in the Union army.

Bickmore's parents insisted that he postpone the dream of a New York museum and earn a reputation as an exploring naturalist. It was decided that a leisurely journey to the Orient and the South Seas was just what the situation required. Arrangements were hurriedly made, and Bickmore was hustled off to Singapore. He wandered through the East Indies, Asia, and Europe until the late fall of 1867, amassing trunks full of rare seashells. He wrote a book called *Travels in the East Indian Archipelago,* then did a crash course in operational techniques at museums in London, Edinburgh, Paris, and Berlin.

Back in Boston by October, 1867, he proved as skilled as P. T. Barnum had been at arranging "advance ballyhoo," and secured lecture engagements at Boston's Society of Natural History. His illustrated talks about "The Ainos, or Hairy Men of Yesso," the exhibits of his seashells, and the publication of his book brought him so much publicity that it was fairly simple to persuade the trustees of Madison (soon to be Colgate) University at Hamilton, New York, to hire him as a professor of natural sciences.

Now that he finally had the proper credentials—international traveler, lecturer, author, and professor—twenty-nine-year-old Bickmore assembled the floor-plan drawings he had made for an American museum of natural history and began to solicit funds from both the aristocrats and the *nouveaux riches* of Manhattan. His first success was with the prosperous plate-glass manufacturer, Theodore Roosevelt. Roosevelt provided him with letters of introduction to the soap manufacturer Robert Colgate, the department-store millionaire Alexander T. Stewart, the copper-mine heirs I. N. Phelps and A. G. Phelps Dodge, and the powerful bankers J. P. Morgan, Levi P. Morton, and Morris K. Jesup.

In December, 1868, Bickmore presented a petition signed by eighteen of New York's wealthiest men to Andrew H. Green, comptroller of Central Park. The petition asked if the city government was "disposed to provide for [the] reception and development" of a "great Museum of Natural History . . . in Central Park." On January 19, 1869, Green replied that "the Commissioners will use their best exertions toward the establishment of a Museum of Natural History."

But Bickmore and his sponsors knew that Green was playing

a political version of the "shell game." Waterhouse Hawkins, the British artist who had designed the Crystal Palace "antediluvian monsters" in 1854, had arrived in New York in the spring of 1868. He had studied the files of Silliman's magazine and read Hayden's articles about the fossil treasures of the Dakota Badlands and Montana's Judith valley. And he knew that Joseph Leidy was preparing a monumental monograph, *The Extinct Mammalian Fauna of Dakota and Nebraska*, for publication. It seemed certain, then, that a virtual "mother lode" of dinosaurs and prehistoric fossils was about to be uncovered in America. So Hawkins proposed to Green that the Central Park Commission lobby for municipal financing of a "Paleozoic Museum," to be built on the west side of the park near the Sixty-third Street entrance. The structure, he urged, should be a replica of London's Crystal Palace. Beneath its glassed central dome, he would construct diorama models of "the great group of ancient animals formerly living during the secondary geological epoch on the continent of America."

Green approved the plan and drummed up enough political support to hire architects and builders and give Hawkins a drawing account. By the time Bickmore and his backers presented the petition for the American Museum of Natural History, Hawkins was in Philadelphia studying Joseph Leidy's manuscript and the Academy's exhibits.

Bickmore's sponsors, however, learned that Green and his associates were battling with "Boss" Tweed and the Tammany Hall ring. Samuel J. Tilden, the attorney who would be the Democratic presidential nominee in 1876 and who, as state Democratic chairman, was already gathering evidence against Tweed, gave Bickmore a letter of introduction to Tweed.

Tweed received Bickmore at his suite in an Albany hotel one February afternoon in 1869. He scanned the letters from Tilden and several of the museum sponsors, then lolled back in an armchair as, with only grunts and nods, he listened to Bickmore talk about the "great public benefits" of the museum. When Bickmore's plea ended, Tweed carefully picked up the letters, stuffed them into a coat pocket, and rose. "All right, my young friend," he boomed. "I will see your bill safely through." They shook hands and Tweed led the way to the door.

On April 6, Governor John T. Hoffman signed the bill granting the American Museum of Natural History a charter of incorporation. A few weeks later, a reorganization of the municipal government did away with the Central Park Commission, and Waterhouse Hawkins's agreement was declared invalid. The hole that had been dug for the foundation of the Palezoic Museum was filled in. The casts of "great groups of ancient animals" Hawkins had begun were used for groundfill near the pond at the Fifty-ninth Street and Fifth Avenue entrance to Central Park. If the charter granted to the American Museum of Natural History had anything to do with the tossing out of both the Central Park Commission and the Paleozoic Museum the details failed to reach the public.

The first public exhibit at the American Museum of Natural History was held during the spring of 1871. It consisted of stuffed birds, mammals, fish, reptiles, and bugs, including "1 bat, 12 mice, 1 turtle, 1 red squirrel skull, and 4 birds eggs" contributed by Master Theodore Roosevelt, Jr., aged nine. The temporary home was a municipally owned building at Central Park and Sixty-fourth Street, across the street from the new grass patch that was to have been the site of Hawkins's museum. By that time, Bickmore had accepted the position of museum superintendent, had resigned from Madison University, and was involved in the construction plans for the Victorian Gothic structure that was to be the institution's first permanent home, in the goat-pasture hills and shanty litter of West Seventy-ninth Street.

The cornerstone of the Seventy-ninth Street building was laid on June 2, 1874, by President U. S. Grant. A year later, the first long step toward resolving the challenges of paleontological exhibits was taken when Bickmore persuaded the trustees to pay James Hall $65,000 for the choicest invertebrate fossils that Hall had acquired since his first trips to the

Schoharie cliffs with Amos Eaton. (Characteristically, Hall carried on an irascible correspondence with Bickmore and his associates for decades, alleging that some of the fossils had been shipped "in error" or were needed in Albany for "detailed study.")

The intricate work of transforming agatized remnants from "pretty rocks" to accurate displays and synthetic replicas of their dawn-age environments was the challenge faced by museums in the final decades of the nineteenth century. The funds necessary for field expeditions, artists, metallurgists, chemists, curators, architects, construction, and transportation could not have been realized by admission fees or memberships. Nor would the political bias against Darwinism, "birds with teeth," and "Devil's tools" allow more than meager government grants to these dawnseeker institutions.

Consequently, the technique that Albert Bickmore and his millionaire hobbyists pioneered in founding the American Museum of Natural History became the standard pattern for American museum development and research financing. The basic team consisted of a dedicated promoter, a few millionaires eager to enhance the family name, and politicians eager for bureaucratic power, construction contracts, a realty boom in the area where the museum would be built, and the status and increased tourist activity that would result from offering people serious "Culture."

The network of land-grant colleges was also being organized during the 1870's. Their chancellors and trustees deftly used the same methods to acquire funds for dormitories, football stadiums, libraries, laboratories, student unions, and "research grants." (Ratification of the Sixteenth Amendment to the Constitution, on February 3, 1913, gave Congress the "power to lay and collect taxes on income." This prompted lawyers, whose mission was to avoid "confiscatory" taxation for their clients, to adopt the pattern of setting up foundations and trusts as "philanthropies for the public good." Many of the foundations—Rockefeller, Guggenheim, Pew, Bush, Surdna, Ford, and others—had a built-in connection with natural history since the fortunes behind them had been amassed through applied-science techniques of processing and marketing fossil materials.

Development of these fund-raising methods was as timely to the dawnseekers as William Bessemer's new process for producing steel was for the nation's industries. The first transcontinental railway had been completed during the summer of 1869. Now the fossil treasures of the Badlands and the Judith could rumble east on boxcars. And Hayden, Marsh, and Cope were already out there digging, learning, and feuding.

14 | High-Iron Treasure

At the end of a long day's march, one of the soldiers, hot, thirsty and utterly weary, was heard to exclaim: "What did God Almighty make such a country for?" To which one of his companions made the reply that "God Almighty made the country good enough, but it's this infernal geology that the professor talks about that has spoiled it all."—George Bird Grinnell, *An Old Time Bone Hunt*

THE transcontinental railroad was a time machine. It shrank the width of the nation from thirty to seven days, brought great regions of high plain and mountain out of the Stone Age and into the Steel Age, and uncurtained the Earth's most awesome panorama of prehistoric flora and fauna.

The geologist was an essential member of each team of surveyors and location engineers that worked fifty to one hundred miles ahead of the graders, track layers, and gandy dancers on the Central Pacific, Union Pacific, and Santa Fe. Veins of iron, copper, coal, sulphur, limestone, marble, mica, and magnesium assured the rise of industries that would be responsible for the success or failure of the decision to adapt this "Great American Desert" to machine civilization. Also, since state and federal departments of Agriculture were still little more than modest agencies for compiling statistics and handing out packets of test seed, the geologists on railroad payrolls scouted for the likeliest farming land, range pasture, water supplies, and other assets that would encourage realty sales.

Scientific journals were eager for tidbits about "the wonders of the West." Skilled journalists, such as Henry Stanley of *The New York Herald*, were sent out to report about railhead construction camps. The discovery of giant shells and bones in tunnel bores, and the Union Pacific's dig through Fossil Fish Cut in the shale hills west of Green River, Wyoming, made headlines in the East and prompted scores of universities and museums to set aside money for paleontological expeditions in the West.

But Joseph Leidy and Ferdinand Hayden preceded them. The Civil War had restricted Leidy's quest for fossils to shell-and-tooth-studded cliffs along the Delaware coast and bone piles in marl pits near Haddonfield, New Jersey, a Philadelphia suburb. In the fall of 1858, a paleontology enthusiast had unearthed pelvis and leg bones, vertebrae, jaw fragments, and nine teeth from Cretaceous marls on the farm of John E. Hopkins in Haddonfield. Leidy identified them as remnants of a dinosaur, closely related to Mantell's *Iguanodon*. He named the species *Hadrosaurus*, in honor of the village, and, in 1871— after the abandonment of the Waterhouse Hawkins's Paleozoic Museum—permitted Hawkins to make a casting to be exhibited at the Academy of Natural Sciences. The Haddonfield marls and Delaware cliffs provided some exercise for Leidy's curiosity about the evolution of America's life forms, but he devoted most spare hours to working on the manuscript of *The Extinct Mammalian Fauna of Dakota and Nebraska*, based largely on the fossils that Hayden had sent him.

The addition to the Academy's staff who seemed most promising to Leidy was the exuberant and often reckless young anatomist, Edward Drinker Cope. Born into a wealthy Quaker family in 1840, Cope became obsessed with natural history at the age of six, when he began, on his own, to explore the dioramas in the Peale Museum. As an eight-year-old at a Quaker day school, he routinely doodled scale drawings of fossils and contemporary flora and fauna in the margins of his notebook. At eighteen, he established a reputation at the Academy of Natural Sciences when Leidy and associates accepted the youth's paper, "The Primary Divisions of the Salamandridae with Descriptions of the New Species," for publication in the spring, 1859, issue of the *Proceedings*.

A few older members of the Academy questioned the wisdom of accepting the work of a mere stripling for the *Proceedings*. But Cope had been on a first-name basis with most of the Academy's staff for years. His genius in plant and animal identification, the accuracy of his drawings, and the scope of his imagination eventually caused Leidy to make discreet inquiries about his record at Westtown School. Family prosperity and Edward's obvious brilliance and capacity for grasping the complexities of natural history counterbalanced his moodiness, his readiness to engage in fist fights, and his "lone wolf" behavior. The youth might, Leidy concluded, develop into a superb diagnostician. And there would certainly be need for such skills as both Darwinism and the opening up of the paleontological wonderland in the West focused scientific as well as lay interest on fossils, the real age of the Earth, and valid evidence of when *Homo sapiens* began. The tediousness of university study could prove deadly for a youngster with such rare potentials. Edward Cope was ready for advanced studies at the Smithsonian and then abroad.

In 1860, Cope enrolled in Leidy's course in comparative anatomy and, during the same months, recatalogued the Academy's herpetological collection. That fall, he wrote to Joseph Henry and Spencer Baird at the Smithsonian asking for permission to become a "special researcher" there. Permission was granted. Cope worked so diligently and initiated so many abstruse discussions that within a year he had thirty-one papers on zoological classifications accepted by the *Proceedings*, "Silliman's Journal," and other natural-history publications. A frustrated courtship in 1862 led to moodiness, retreat to a farm his father had given him in New Hampshire's White Mountains, and months of introspection about Darwinism. (He eventually became a convert to and a leading American exponent of Lamarck's theory of acquired characters.)

In 1863 and 1864, Cope took Leidy's advice about study abroad and went to Europe. In Berlin, Paris, and London, Cope intrigued museum curators and diagnosticians as promptly as he had won the co-operation of the Smithsonian staff. He returned to Philadelphia in the fall of 1864, showed his notes and sketches to Leidy, and convinced the trustees of Haverford, a Quaker college in Pennsylvania, of his qualifications for the post of professor of zoology.

Leidy permitted himself a sigh of satisfaction. The tenacity of U. S. Grant and William Sherman made it seem more and more likely that Lee's army and the Confederacy would crumble within the year. Cope was home again and seemingly ready for the challenge of the Far West. Dr. Hayden had volunteered as a surgeon in 1862 and was now a lieutenant colonel. If he survived the war, an expedition might—just barely might—get into the field by 1866.

Colonel Hayden was mustered out of the Twenty-second Division of the cavalry in the spring of 1865. Leidy and other friends at the Academy had secured a university appointment for him—Hayden was offered the chair of geology and mineralogy at the University of Pennsylvania, with the special proviso that university routines would "not interfere with Western expeditions."

Plans for a return to the Badlands began to take shape that fall. Leidy needed more environmental data to resolve questions that kept arising out of the *Mammalian Fauna* manuscript. Hayden felt it important to study strata and fossil beds along the Niobrara River, west of Fort Randall and across the sand hills to the south end of the Badlands. The Academy would finance the operation for the summer months of 1866. There would be enough funds also to hire the lank Kentuckian James Stevenson, a veteran of the 1857 and 1859 expeditions.

Hayden and Stevenson reached Fort Randall in July, rented a 1,775-pound wagon and a six-mule team, and were assigned an escort of four army privates, a corporal, and two Indian guides. The 660-mile circuit up the Niobrara, northwest to the foothills of the Black Hills, east to Fort Pierre, then across the Missouri's loop to the mouth of White River, took six weeks.

The most puzzling packet of bones among those accumulating on the wagon added mystery to the Sioux legends about "ghost horses." Sparkling at the base of one of the Badlands turrets, the skeleton looked at first glance like the remains of a sheepdog. But the muzzle, teeth, hooves, and rib cage all identified it as a tiny horse. However, its hooves had three toes. Here, if the Leidy-Cope diagnosis agreed with Hayden's, was startling testimony to Darwin's theory of the evolution of a species. Other bundles on the wagon contained bones strangely like those of camels and the first fossil insectivores ever found in the West.

Leidy and Cope agreed that the collie-sized skeleton was that of a three-toed horse, and that the other unfamiliar bones were those of an ancestor of the camel. Leidy was certain that scores—perhaps hundreds—of unknown species could be exhumed from the Badlands and the Judith country. If this was true of these areas, what else was hidden in that wild sprawl of strata between the Missouri, the Pacific, Canada, and Mexico?

Spencer Baird of the Smithsonian was certain that the essential first step in learning about those stratal mysteries and their epochal contortions was to institute federal geological surveys. The geologists hired by the railroads were, after all, concerned only with tunnels, rights of way, and the economic potentials of the belt of land Congress was granting to the railroad builders. Political shenanigans would undoubtedly limit the value of any surveys undertaken by local governments. Now Nebraska was about to achieve statehood. Her vast prairie, sand hills, and buttes demanded a geological survey conducted by a professional with trans-Missouri experience. The obvious choice was Professor Hayden.

General John A. "Black Jack" Logan agreed. A freshman congressman from Illinois, Logan was national commander of the Grand Army of the Republic (GAR). Chicago was bidding for the economic domination of the West by pushing through a rail connection with Union Pacific's eastern railhead at Omaha. In response to home pressures and the war veterans' demands for homesteader grants in the West, Logan became Congress's most outspoken advocate of economic exploitation of those high-iron treasures.

Baird was already assisting Clarence King, a veteran of the California Geological Survey, to obtain congressional approval of a geological survey along the 40th Parallel from western Wyoming to the California border. General Logan proposed, and won, a five-thousand-dollar appropriation for a survey of

Nebraska, led by Hayden, "to be prosecuted under the direction of the Commissioner of the General Land Office." The ambitions of GAR and Chicago lobbyists broadened Hayden's instructions. By the spring of 1867, he was instructed not only to "make ample collections in geology, mineralogy and paleontology to illustrate the notes taken in the field," but also to identify "beds, veins, deposits of ores, coals, clays, marls, peat and such other mineral substances as well as the fossil remains of the various formations." In addition, he was to examine "soils and subsoils, describe their adaptability to particular crops and the best methods of preserving and increasing their fertility," and "look into the feasibility of introducing types of forest growth." And, since five thousand dollars was a lot of money for a rock hunt, certain members of the House Appropriations Committee—all devout church deacons—insisted that there be some concise "pictures and the like in that fellow's Report."

The appropriation was confirmed by late April, 1867. Leidy and Hayden devoted afternoons and evenings to plotting the expedition's itinerary. They were discussing the prospects for fossils and minerals on the sand-hill approach to the Badlands one June afternoon when the office door was flung open and Professor Edward Drinker Cope of Haverford stalked through. Cope slumped into a chair and, in the sepulchral tone of a tragedian rehearsing a scene from *Hamlet*, groaned, "I despair of that place! Flummery there is at Haverford." After much questioning by Hayden and Leidy, Cope finally told them that he had resigned from Haverford and was about to take his bride to their New Hampshire farm to think through whether he really should purchase a home on the edge of those Haddonfield marl beds and devote all of his energies to "worth-while research." Leidy nodded his understanding, but he permitted his left eyebrow to tilt slightly as his eyes met Hayden's. "Now," his expression clearly said, "Nebraska's fossils have acquired a cracking good diagnostician."

Gifts of scientific tools and supplies from the Smithsonian, the loan of mule and horse teams from the army quartermaster at Omaha, and railroad passes and the promise of "all kinds of co-operation" from Union Pacific made it possible for Hayden to outfit his most elaborate expedition. James Hall gave Fielding Meek a summer's leave from Albany to serve as Hayden's "professional assistant." Jim Stevenson and three young graduate students who would serve as "specimen collectors" completed the team. The cavalry patrols and the daily clatter of supply trains between Omaha and the railhead made the Platte valley route seem almost as safe as the University of Pennsylvania campus. (But it wasn't. That August 6, Chief Turkey Foot and 150 Cheyenne cut the telegraph line at Plum Creek, 60 miles east of North Platte, piled rocks and brush across the track, then sat down to watch an Iron Horse take a pratfall. The yield was three scalps, a spectacular fire, and so much tempting plunder from the boxcars that Major North and his Pawnee Scouts easily caught up with the Cheyenne two days later, and killed fifteen of them.)

The Casement brothers' "Hell on Wheels" construction train was laying iron at a two-miles-a-day clip up Lodgepole Creek toward the rowdy new tent town of Cheyenne. The Hayden brigade serenely crisscrossed Nebraska south and north of the twenty villages mushrooming alongside the railroad. Probes for water attested that the best hope for Nebraska agriculture would be deep wells powered by those gawky, galvanized windmills that had just been invented by a man named Wheeler in Beloit, Wisconsin. The prospects for fuel were grimmer. No evidence of anthracite or even lignite showed in surface strata or test bores. The cottonwoods that dominated creek valleys were almost as useless for chunkwood as they were for railroad ties. Nebraska's fuel might have to come "over the hump" from Wyoming, or perhaps Colorado. The sod houses that some villagers were building seemed to offer the best assurance of maximum winter warmth with minimum fuel.

At Julesburg, where the Casements had recently shipped scores of prostitutes east in an effort to check the plague of venereal disease among construction crews, the survey and its cavalry escort trekked up the South Platte toward Denver.

The test bores Hayden and Meek made between Fort Collins and Denver were far more promising for industry in the West. The coal and lignite beds seemed extensive, and there was iron too. If those gold and silver digs on the Front Range proved successful, Denver could become another Pittsburgh.

The samples of fossils and strata that Meek took back to Albany to show James Hall were so exciting that Hall decided to ride along with Hayden if Congress renewed the five-thousand-dollar appropriation for 1868. Moreover, Hall had been corresponding with Clarence King. Hall and Meek might take on the paleontological analyses for the 40th Parallel survey and thus help pioneer fossil beds on the Pacific slope.

From "Silliman's Journal" that winter, Hayden learned that his brigade had been within a two-days ride of a group that Professor John Wesley Powell had led into the Pike's Peak region. A professor of geology at Illinois Wesleyan University, Powell also had solicited funds for a geological survey in Washington during the spring of 1867. He wanted to map and study the sprawl of mountains and gorges in western Colorado where the Colorado River and the Rio Grande rose. But he had lost his left arm at the Battle of Shiloh in 1862 and this handicap, despite his valiant record and his promotion to major, prejudiced officials against him. So he had been brushed off with a token appropriation of a few army mules and some horses and supplies. Yet he had organized a squad of students and friends into an assault on Pike's Peak, a hundred miles out of Denver. Defiantly, Major Powell led the way up the peak. His wife was in the party too, thus becoming the second woman ever to ascend the peak. In 1868, Powell made it plain that he intended to go farther west, into those "dark lands" of the Gunnison and perhaps clear across to the valley of the Green.

The first annual report of the Nebraska survey ran to only sixty pages. But it was succinct and so practical and promising in its details about the potentials for industry and agriculture that General Logan and Spencer Baird were confident that another five thousand dollars could be squeezed into the sundry civil expenses budget for 1868. Hayden passed this news on to

James Hall. When classroom duties permitted, Hayden pored over charts and maps with Leidy and Cope to decide on the likeliest places for paleontological digs and mineral veins in northwest Nebraska and Wyoming. There was evidence of Cretaceous strata behind Denver; there must be dinosaur, mosasaur, and other marine fossils hidden in that grandeur.

Cope was taking on enough projects to keep six diagnosticians occupied, and so was happier than he had been since those first "honeymoon weeks" at Haverford. Both Ohio and North Carolina wanted him to analyze the strata and fossil samples their state geological surveys had uncovered. This was sufficient excuse to take his wife and baby daughter, Julia, on camping trips so that he could examine strata, explore caves and Indian mounds, and make long forays to search out unusual flora and fauna. Also, he had ferreted six more dinosaur specimens out of the Haddonfield marls. Now, with much grumbling about the "Irish shanties" and the "filth" and "excessive smoke" of Philadelphia, he was buying that home at Haddonfield, with enough land for gardens and a retreat where he could research and write in peace.

In mid-March, 1868, Cope bustled into Hayden's office with a scowling fat man who looked like a mirror image of Queen Victoria's beer-bellied, bearded son, Edward. Hayden rose and smiled politely when Cope introduced his companion as Professor Othniel Marsh of Yale. "Silliman's Journal" had joyfully announced in June, 1866, that George Peabody had given Yale $150,000 for the Peabody Museum of Natural History. There was little surprise among natural historians when, on July 26, 1866, Yale also announced the appointment of Peabody's nephew, Othniel Marsh, as America's first professor of paleontology.

Marsh and Cope, it seemed, had met while they were studying in Berlin in 1863. Marsh had come to Haddonfield to explore those lovely marl beds and to examine Cope's collection of bones, which he believed to be from plesiosaurs and mosasaurs, marine monsters of the Cretaceous seas. Marsh and Cope would explore the marl beds for a week and discuss the bones.

Marsh had just heard that complete mosasaur skeletons were being dug up along the right of way of the Kansas Pacific Railroad. He planned to go out there and examine them.

There was brief, formal discussion about Hayden's beloved treasure troves in the Badlands and upper Missouri–Judith region, with blunt questions about Hayden's proposed route that summer "in event your budget is approved." Then, with a curt farewell nod, Professor Marsh went off to keep an appointment with Joseph Leidy.

Washington was so caught up in President Andrew Johnson's feud with Secretary of War Stanton and Johnson's two-month impeachment trial that the sundry civil services budget was not approved until July, 1868. But reassurances from both the Smithsonian and Congress, plus specific orders to "concentrate on Wyoming," prompted Hayden, James Hall, and James Stevenson to meet in Omaha in late June.

It seemed certain that Congress would soon approve the bill creating the Territory of Wyoming. Then Cheyenne, its redlight district recently cleaned up by a vigilante committee, would become territorial capital. In April, Hell on Wheels completed the track and teetering wooden trestles over that Mesozoic-Paleozoic-Precambrian "ladder" at Lone Pine Pass, paused for girls and booze at newly established Laramie, then clanged on to the bleakness of Big Basin.

Hayden's choice of Fort Sanders, the army post guarding Laramie and the west end of Lone Pine Pass, as base camp for the 1868 survey was not explained in the year's report. The order of events suggests his reason, and underscores his political deftness.

On May 21, 1868, the National Republican Party nominated Ulysses S. Grant as its candidate for president of the United States. On July 9, the Democrats nominated New York's exgovernor, Horatio Seymour, as its nominee for president. Major General Grenville Dodge had served spectacularly since 1865 as chief engineer of Union Pacific. The two "scorpions under the bedsheet," as Jack Casement referred to them, were Thomas C. Durant, Union Pacific's shrewd general manager,

and Silas Seymour, a consulting engineer. Silas was Horatio
Seymour's brother.

The bitterness between Durant and Dodge climaxed at a confrontation in Laramie in June. Thereafter, Durant vowed that
he would fire Dodge and appoint Silas Seymour chief engineer.
The boast was relayed to Washington and to General Grant. In
mid-July, Generals Grant, Sherman, Sheridan, and Harney
were scheduled to arbitrate with the Sioux, Cheyenne, and other
Indian nations at Fort Laramie. Grant bluntly requested that
Durant, Seymour, Dodge, and "appropriate" Union Pacific
directors be at Fort Sanders on July 26 for "a review of the
charges against General Dodge." The conference lasted two
hours. At its close, Grant leaned forward and rasped, "The
government expects the railroad company to meet its obligations . . . and the government expects General Dodge to remain with the road as its chief engineer."

There is no positive evidence that Hayden had an audience
with General Grant at Fort Sanders. But it is true that General
John Gibbon, army commander of the Rocky Mountain District, personally escorted Hayden and Hall on a rock hunt and
mine inspection in the Snowy Mountains, southwest of Laramie,
and that Hayden did join General Dodge and General Francis
Preston Blair on an antelope, elk, and mountain goat hunt to
Colorado's North Park, south of Fort Sanders.

General Dodge and Hayden had much to talk about. Dodge
had learned surveying as a fourteen-year-old in South Danvers,
Massachusetts. In the 1850's, Dodge had benchmarked canal
and railroad routes from the Chicago River to the Platte. He
had been General Grant's special adviser on the rebuilding of
the railroads the Confederates had wrecked during their 1862–
63 retreats down the Mississippi valley. Perhaps it was Dodge
who first told Hayden, beside a North Park campfire, about that
Fossil Fish Cut the blasters had exposed while preparing the
grade just west of Green River City, and about the abundant
evidence of ancient life gleaming in the washouts of Bridger's
Basin. In any case, the North Park hunt ripened a friendship

that gave Hayden access to special consideration by both the White House and the War Department, if and when he needed it.

In late August, the two-horse ambulance, prairie schooner, and saddle horses of the survey crossed the Great Basin toward the Green River and Fort Bridger. The stumps of palm trees, agatized to pastel blues and tans, gleamed in the valley of the Medicine Bow River. Eocene bones and shells and skulls fretted the rocks around Cooper Lake. On the Bitter Creek approach to the Green River bridge they passed Hell on Wheels, now laying high iron at a six-mile-a-day rate. Hayden made his first examination of the Fossil Fish Cut, identified the rock as Eocene, named it the Green River Formation, and selected samples of its bass, herring, and sandfish to take east to Leidy and Cope. (Hayden's validation of the antiquity of Fossil Fish Cut specimens and subsequent analyses by Cope and Marsh instigated a rash of pilferage that continued for decades. Several homesteaders made lifelong careers of blasting and sawing fish fossils out of the cut and from other fossil fish beds in the vicinity. The Green River City depot became their favorite sales center. "Tourist-car passengers had a choice of the common *Knightias* for $1.50 to $3.00, or Sunfish for slightly more," Gilbert F. Stucker reported. "Holders of Pullman tickets were shown the larger, deep-bodied varieties which fetched $15 to $35, while stateroom occupants were offered the 'museum pieces' valued at $100.")

The aspens were deep golden when Hayden routed the expedition south through Denver to Fort Garland, Colorado, conducting soil and mineralogic samplings along the way and selecting rock and plant samples for Leidy, Cope, and the Smithsonian. From Garland, in October, he made a rapid, and consequently misleading, survey of the ranch lands recently purchased by an English friend, William Bickmore, in the former Apache heartland, the Sangre de Cristo Mountains. He did not return to Philadelphia until early November, long after the fall semester had begun at the University of Pennsylvania.

If Hayden's "socializing" out of Fort Sanders was really just shrewd politicking, it stemmed from his zealous dedication to the future of geological and paleontological awareness in and about the West. And the reward was prompt. Grant was elected by an electoral-vote landslide of 214 to 80, but a popular majority of only 309,594. Within two months, the Smithsonian's lobbyists had assurance that the administration of Hayden's survey was being transferred from the General Land Office to the secretary of the interior, its title was being broadened to the United States Geological Survey of the Territories, and the annual budget would probably be doubled to ten thousand dollars "for the time being."

Joseph Leidy watched the pattern unfolding with intense interest. The hunt in the West was finally under way. Clarence King, with his topographers and naturalists, had researched brilliantly through the Wasatch Mountains to the Great Salt Lake basin; during 1869 they planned to measure, analyze, and collect samples from those bleak expanses across western Utah and Nevada.

If Hayden and Generals Blair and Dodge had ridden out on the North Park hunt two weeks earlier and had trained a telescope on Long's Peak, they could have seen John Wesley Powell and his companions inching up that huge granite capstone to the summit. Now, it was said, Major Powell, his wife, and seven adventurers were wintering in the Green River wilderness and making preparations for a daredevil float down the Green, the Colorado, and through Grand Canyon in 1869.

In the same late-summer weeks, Professor Othniel Marsh came west to Laramie and made numerous inquiries about routes and military escorts for "a scientific expedition" to the Badlands. He seemed to have a suitcase filled with letters of introduction to fort commanders, railroad officials, and politicians. At Antelope Station, Nebraska, one of the Union Pacific's geologists gleefully reported, well diggers had dredged out buckets full of bones and skull pieces. Somebody, doubtlessly a devout fundamentalist, started the story that the bones were fragments of "one of those Biblical giants." The conductor of Marsh's train, seeing an opportunity to put Antelope Station on the map, repeated the remark to his distinguished passenger. Marsh bribed the conductor to hold the train at the station long enough to permit a hurried examination of the pile of earth. Then he made a deal with the stationmaster to have all the bones from the pile packaged for him by the time he returned east. "He got a hatful," Leidy's informant wrote, "and seemed pleased as punch about it. He bragged that they were parts of a prehistoric horse."

And now Cope seemed ready to ride west, too. He was as prolific as ever—he had written twenty-six scientific papers during 1868. The fossils sent him by Hayden and the boxes of bones and flora from the Santa Fe and Smoky Hill railroad digs made him eager to finish the North Carolina and Ohio analyses and get into that great hunt in the West.

King and Powell and Hayden and Marsh and Cope were all determined men, eager men, haughty men. Science, industry, this noisy maze called "American civilization," would profit by their intensity and their skills. But there would be power struggles too. Leidy sighed and recalled Herbert Spencer's apt phrase: *survival of the fittest.*

15 | How to Find
a Heritage

The early fossil hunters surely had to endure hardships in the days when West meant West. I can well remember how most of my western friends regarded the early fossil hunters and naturalists who came to do collecting. They were usually spoken of as bone- or bug-hunting idiots. For anyone to go chasing over the West hunting for petrified bones, or even bugs, was conclusive evidence of his lack of good horse sense, especially in sections of the West where Indians were still wild enough to want to stick their arrows into anything wearing a white skin.—James H. Cook, *Fifty Years on the Old Frontier*, 1923

OTHNIEL MARSH quickly and confidently identified the agatized splinters projecting from the dig heap of a Nebraska well as rare, and possibly still unique, remains of a prehistoric horse. His ability to do so confirms three conclusions relevant to the paleontological achievements of the 1870's.

First, Colonel Jewett's seminars at the Lockport gorge had obviously been thorough. Second, Marsh had inherited the Peabody tenacity and passion for detail that had enabled his uncle George to parlay a Georgetown, D.C., dry-goods shop into one of the world's most powerful banking firms. (The house of Morgan grew out of Peabody & Morgan.) Third, the 1780–1860 researches of Cuvier, Wistar, Eaton, Silliman, Hall, Buckland, Lyell, and Leidy established enough benchmarks for paleontology to enable their successors to know what they were looking at.

Bulldog tenacity, know-how, and a zest for continuing education through research were the three qualities that enabled a few squads of "bone- and bug-hunting idiots" to pioneer and popularize panoramas of Western heritage that still awe mankind and stimulate its curiosity.

Marsh was thirty-seven years old in 1868, only two years younger than West-wise Hayden and nine years older than Cope. The nickname of "Daddy," tagged on him by fellow students at Phillips Andover Academy, the dourness of his father, and the difficult years on the Lockport farm all contributed to Marsh's reputation as a haughty and ruthless schemer. The passion for paleontology that was born during the Lockport gorge afternoons with Colonel Jewett evoked scorn and abuse from Marsh's father, who complained that his son was "wasting good money" to take courses at nearby academies. The courses, however, qualified Marsh for a teaching job in a rural school when he was twenty. The next year he came of age and inherited twelve hundred dollars from his mother's share of some Peabody property. He used this to enroll at Phillips Andover; most of the other freshmen were twelve to fourteen years old.

The taunts of his fellow students, the death during childbirth of his favorite sister, and the challenges presented by Phillips Andover teachers combined to make Marsh the school's most intense student—and politician. He stood first in class every term, maneuvered his election to the presidency of the debating society, and captained the football team.

Meanwhile, Uncle George's compassion for "the poor and uneducated" was receiving so much publicity that there was a movement to change the name of his Massachusetts Bay birthplace from South Danvers to Peabody. Baltimore, where the transition from dry goods to banking had been made, received a $1,500,000 gift for a Peabody Institute. Another $2,000,000 was parceled out to British institutions to improve "housing for the poor." Queen Victoria sent word that a baronetcy could be added to the next honors list; Mr. Peabody graciously thanked Her Majesty and pointed out that he was an American and so should not accept a title.

Assured that he would be the class valedictorian, Othniel sent Uncle George an invitation to the 1856 graduation at Phillips Andover, recited his achievements, and expressed an urge to continue his pursuit of knowledge at Yale University. The prospect of a scholar in the family so pleased Uncle George that he promised his nephew tuition plus a hundred dollars a month *if* Othniel could pass the Yale entrance examinations.

By October, 1856, when millionaire and nephew finally met in Washington, D.C., Marsh was a Yale freshman with a four-room apartment that was rapidly being filled with New England and Nova Scotia fossils and rock samples. The following July, Peabody paid his first visit to Yale, questioned Othniel's instructors, and dined at Benjamin Silliman's home. Professor Silliman was seventy-eight that year, but he was far from senile. In his mind, he added Peabody and Marsh and came up with a rosy dream: the Peabody Museum of Natural History.

After that, Marsh became a carefully coached student. He was elected to Phi Beta Kappa and ranked eighth in the 1860 graduating class; Silliman saw to it that he was awarded a Berkeley Fellowship, which provided tuition and room and board for graduate studies. Uncle George promptly sent a check that would enable Marsh to afford a larger apartment and field trips during his two years at Sheffield Scientific School.

The outbreak of the Civil War imperiled Silliman's dream. Othniel Marsh was a belligerent Unionist, a devotee of hunting and hiking, and, as his leadership skills at Phillips Andover indicated, a logical candidate for a regimental officer. He was nominated for major of a Connecticut regiment, but his eyesight was so poor that reading glasses were essential, and the risk of breakage would be too great, even for that quick march into Virginia that would most certainly vanquish the Rebels. Marsh returned to the studies of chemical geology and mineralogy and by 1862 knew as much about both subjects as the faculty.

Benjamin Silliman was the first to suggest that Marsh do further graduate work in Europe. Uncle George agreed, and he provided the funds to back up his sentiments. Soon after his thirty-first birthday, Marsh sailed for Germany and became a student assistant to Ferdinand Roemer, a famous paleontologist at the University of Breslau. That winter, after long discussions with Uncle George—while the millionaire soothed his gout in the waters of Wiesbaden—the Silliman dream materialized. Marsh wrote that he had obtained a promise of $100,000 for a museum of natural history at Yale. What studies, he asked, would be most likely to lead to consideration for an appointment to the Yale faculty?

Silliman replied that, since it would have a natural history museum, Yale should be the first American university with a professorship in paleontology. During 1864 and most of 1865, Marsh commuted between Berlin, London, and Paris asking questions, taking notes, collecting books, and using the prestige of both the Peabody name and Yale University to obtain interviews with Darwin, Lyell, Huxley, and curators and keepers of major museums. In the process, he became an outspoken advocate of Darwinism, which later led to denunciations from fundamentalist congressmen as well as Edward Drinker Cope.

Back at Yale by the fall of 1865, Marsh rented a three-room apartment, took his library and fossil collection out of storage, elaborately cross-indexed them with the trunks of specimens he had brought from Europe, and began a program of visits to and evaluations of natural history museums between Montreal and Washington. In the late spring of 1866, Yale received $150,000 in bank drafts and bonds from George Peabody for founding the Peabody Museum of Natural History.

There was a verbal agreement with the Yale trustees. On July 24, Yale announced its creation of the nation's first chair of paleontology, with the professorship going to Othniel Marsh. There would be no teaching duties—or salary; Peabody and Marsh had already agreed on a bequest to cover Marsh's living expenses in New Haven, plus a drawing account for travel and research. (Somewhat put off, perhaps, by his nephew's blunt espousal of Darwinism, Peabody also gave a gift to Harvard University to establish a museum dedicated to anthropology. The Harvard and Yale endowments may have influenced Albert Bickmore's 1868–69 campaign to solicit funds from New York City's millionaires for the American Museum of Natural History.) Through drive, ambition, an unrelenting capacity for working with detail, and social deftness, Othniel Marsh ob-

tained the knowledge that enabled him, during a five-minute
train stop, to identify the dawn-horse fossils protruding from
the dirt heap at Antelope Creek, Nebraska. Analyses of the
bones at Yale that winter convinced him that his "new fossil
treasures" were skeletal parts of "horses, diminutive indeed,
but true equine ancestors. . . . The small horse was strongly in
evidence. . . . During life he was scarcely a yard in height,
and each of his slender legs was terminated by three toes. Later
researches proved him to be a veritable missing link in the
genealogy of the modern horse." (Hayden had discovered the
skeleton of a similar horse "about the size of a collie dog" in
the Badlands in 1866, but did not investigate its possible
origins. Marsh did investigate, and was instrumental in tracing
the development of the Equidae through sixty million years,
beginning with the appearance of rabbit-sized *Eohippus* in the
swamp jungles of what is now Wyoming.)

Construction planning for the museum and the death of
George Peabody forced Marsh to cancel his plans for a summer-
long hunt in the Dakota Badlands. The processing of his uncle's
will made Marsh, a bachelor, wealthy enough to afford easily
the network of fossil scouts, the digs and digging crews, the
elaborate packaging materials, the artists and taxidermists and
glassblowers and attorneys who during the next twenty years
would establish the Marsh Collections of the Peabody Museum
into the world's most awesome array of dawn-era life.

The budget increase to ten thousand dollars, and an impressive new designation, the U.S. Geological Survey of the Territories, enabled Ferdinand Hayden to organize his first truly
scientific expedition for the 1869 search across Colorado into
New Mexico. In all of the paleontological pioneering he had
done in the West since 1853 he had served as ethnologist,
botanist, geologist, anthropologist, paleontologist, entomologist,
and, occasionally, artist. But by 1869 the United States was so
intrigued by science that scores of specialized professions were
in genesis. Just as Benjamin Silliman and Amos Eaton had
pioneered the teaching of natural history as a means of earning

a livelihood a half century before, the Reconstruction decades saw the birth of many new scientific professions at universities, in government agencies and museums, and in industry. Each of these specialized groups developed its own vocabulary and its own interprofessional feuds. Hayden accepted the trend and signed on a mining engineer, a zoologist, a Lutheran preacher turned entomologist, and an artist. These, plus assistants and cook and wranglers, formed a party of twelve and necessitated a train of two covered wagons, an ambulance, and a *remuda* of eighteen horses and mules.

Again Hayden based at Fort Russell on the Laramie plain and outfitted with army supplies, wagons, mules, and castoff cavalry horses. Political and economic expediencies dictated his route. Colorado's population had increased by fewer than five thousand since 1860. But Union Pacific was building a spur into Denver from its main line at Cheyenne and Kansas Pacific's hammermen were adding as much as two miles of trackage a day to the Smoky Hill roadbed out of Saint Louis. And now, with the bitter memory of seventeen dead in the stagecoach captured and burned by Indians at Cimarron Crossing on the Santa Fe Trail, the Atchison & Topeka Railroad was reorganizing and preparing to get trackage over Raton Pass into Santa Fe. There were demands for more information about the mining and agricultural potentials and the topography of the peaks, hidden valleys, and waterways between Denver and Santa Fe. Between late June and October, Hayden sought scientific answers to these questions in the soils, strata, fossil beds, and benchmark readings from Cheyenne through Denver to Taos and back again.

Historians are still bickering as to whether Hayden's third annual report was directly responsible for General William A. Palmer's decision to build the Denver & Rio Grande Railroad up the Arkansas River's Royal Gorge toward Utah. There is unanimity, however, about the passage that Hayden failed to edit from the Reverend Cyrus Thomas's section of the report, "Agriculture in Colorado." Several times that summer, the sur-

vey wagons had to stop in order to look for fording places
across creeks. The Front Range had experienced extraordinary
rainfalls; record fruit and vegetable crops covered the bottomlands. A Mexican farmer, seeking to flatter the Reverend
Thomas, remarked, "You Americans must bring good rains
with you," and Thomas was apparently eager to believe him.
He formulated a theory stating that "as the population in-
creases, the amount of moisture will increase. This is the plan
which nature herself has pointed out." Fifteen years later,
Thomas's naïveté and Hayden's careless editing were cited as
reasons for denying Hayden the post of director of the U.S.
Geological Survey.

Nevertheless, Leidy and Cope and the Smithsonian were delighted with the boxes of fossils brought in by the expedition,
and industrialists praised the thoroughness of Percifor Frazer's
section of the report, "Mines and Minerals in Colorado." The
sketches made by Henry W. Elliott, on summer leave from his
post as secretary to Joseph Henry at the Smithsonian, were as
useful to realty promoters as they were to geologists and
engineers. The report sold out its first run of 8,000 copies
within three weeks and is credited with stimulating the rush of
health seekers, homesteaders, prospectors, jobbers, and merchants that swelled Colorado's population from 39,864 in 1870
to 194,327 in 1880 and lured another 28,000 over the passes
into New Mexico.

While Hayden's wagons rutted across the future homeland of
Rocky Ford melons, sugar beets, trout farms, Hereford feed
yards, steel mills, the U.S. Air Force Academy, and luxurious
hotels, another momentous journey was being made, into the
"lost world" of the Green River.

The rainbowed immensity of the Colorado River's channel
through northern Arizona inspired Spanish and American explorers to name it the Grand Canyon. It was, all agreed, unique
in the multistrata grandeurs of its two-hundred-mile sweep and
in the grotesque beauty of the adjacent Painted Desert and
Petrified Forest. But no human being had yet dared a trip

through the Grand Canyon or traced its river to its wilderness source in the Colorado or Wyoming Rockies. This was John Wesley Powell's pilgrimage; he embarked on it as ardently as the knights of medieval Europe had set out on the Crusades.

Powell's future as a professor of the natural sciences was imperiled, probably shattered, in October, 1868, when he failed to return to Normal, Illinois, for "professional obligations." After spending the winter in three sod-and-log cabins they built 175 miles southeast of Green River City, Mrs. Powell agreed that her husband and the six others already pledged to the venture could train themselves to survive in the wilderness. In March, 1869, Powell began the second phase of this training program by ordering a march out toward the Union Pacific's new trackage at Green River City. With Mrs. Powell on mule-back, they packed through the junipers to the valley of the Yampa, crossed over to the Green past the canyoned bleakness of the future Dinosaur National Monument, then clambered through Flaming Gorge and crossed the divide to Jim Bridger's fort. By May 10, they were in Green River City, where Mr. and Mrs. Powell boarded an eastbound train.

Mrs. Powell returned to her parents' home in Detroit. Powell stopped off in Chicago long enough to contract for the construction of four boats. Three of them were to be double-ribbed oak, twenty-one feet long, four feet wide, with sixteen-inch draughts, and outfitted with storage bulkheads at each end beneath double prow and stern posts. The fourth craft was to be only sixteen feet long and of white pine.

Solicitations in Washington in early April were almost as fruitless as they had been in 1867. The Smithsonian promised to send some scientific equipment to Green River in exchange for priority on data and whatever fossils and rock samples Powell might "live long enough to collect." The War Department agreed to send surplus flour, jerky, and salt-crusted fat-back over from its Fort Bridger warehouse. But congressmen were in such a flurry about President Grant's cabinet appointees and the imminence of a nationwide celebration in honor of the completion of the transcontinental rail line that few of them would take the time to talk to "that one-armed professor with some crazy scheme about a river somewhere in the West."

The trustees of Illinois Normal and Illinois Industrial universities were far more understanding, despite Powell's disregard for his contract; they contributed grants totaling $1,100 to the Colorado River Expedition. This, plus $2,000 that Powell had saved from his army pay, constituted the financing of the expedition. However, when he returned to Chicago to oversee completion of the boats, Powell learned that the promotion specialists on the payrolls of Chicago & Northwestern and Union Pacific were as interested in encouraging tourism as they were in selling virgin farmlands of Nebraska, Wyoming, and Colorado to settlers. They waived all costs for delivery of boats, supplies, and manpower to Green River, requesting in return only a guess from Powell about possible scenic tours and resort hotel sites down the Green and the Colorado.

Boats, supplies, and men were on the Green's embankment by mid-June. There were nine men besides Powell, including his younger brother Walter, who had never recovered emotionally or physically from the time he had spent as a Confederate prisoner. The most dependable of the crew, Powell felt, were Oramel and Seneca Howland, natives of Vermont and skilled mountaineers. The others were boisterous adventurers and professional hunters who resented Powell's demands for identification of rock strata, rare flowers and plants, and "pewtrified stuff."

The mystery of the sixteen-foot pine boat was resolved when Powell explained that it was to be the advance craft of the flotilla and would carry him, with Jack Sumner and Bill Dunn as boatmen, plus signal flags, three guns, and scientific instruments. It was christened *Emma Dean,* in tribute to Mrs. Powell. The three oak craft, named *Maid of the Canyon, Kitty Clyde's Sister,* and *No-Name,* would each carry a ton of the army supplies and a minimum crew of two.

After two weeks of practice rowing, responses to Powell's gun and flag signals, and disaster rehearsals, Powell decided that Oramel Howland would assist him in drawing maps and that the thirty-three-year-old Yankee George Bradley was to watch for unusual rock formations, especially fossilized bones, plants, and fish. Powell would order a halt, if the rapids permitted, every fifty miles, in order to determine longitude and latitude; he would also take barometer readings at least three times a day.

"They took off," according to the Union Pacific's telegrapher on the afternoon of May 24, "right after lunch today. If all of those army beans get wet and start swelling, it'll dam the river and flood Green River City."

Powell became so absorbed in studying the "barren desolation" and water-sculptured cliffs "suggesting rude but weird statuary" that the *Emma Dean* snagged on a sandbar the second morning out, and two of the other boats shuddered in alongside. There was no apparent damage; the craft were pushed and poled off, but camp was made nearby and a cliff-top hunt yielded a young mountain goat for the party's dinner.

Despite the hypnotic charm of the "flaming, brilliant red gorge . . . surmounted by broad bands of mottled buff and gray" that shadowed the Green's course through the Uinta Mountains, Powell was more vigilant with his signals. The rapids of Flaming Gorge and Kingfisher Canyon were skimmed without mishap. By that time, too, the crew had learned enough geology to realize that sandstone cliffs usually meant "white water" ahead, because the abrasive power of the river, the rip and tear of winter ice, and the push of windstorms had combined to tumble boulders and sludge into the channel. "In the canyon," Oramel Howland recorded, "we would run a bend in the river and prospect ahead with our light boat, and signal the large boats to come on if it was all right."

But disaster struck on June 8, in the Canyon of Lodore. The *No-Name,* with the Howland brothers and Frank Goodman aboard, careened into a rock, bucked its crew into the rapids, veered on, crashed against more boulders, and then, with keel and ribs shattered, grounded on an islet in mid-river. The crew of *Emma Dean* maneuvered out to the shoal, helped the Howlands and the semiconscious Goodman aboard, and poled and rowed safely back to the bank. The next morning, the same crew got to the *No-Name* and rescued Powell's barometer and a two-gallon keg of whisky. But the guns, ammunition, clothing, and bedding in the *No-Name*'s compartments had been washed downstream, her provisions were hopelessly waterlogged, and army beans wriggled up over the stern like a migration of cactus worms.

The *No-Name* and its grubstake were abandoned, and the supplies from the other boats were piled on shore. It took four days to portage the provisions around the rapids and rope and toboggan the empty boats through. "Working like galley slaves all day," George Bradley wrote in his diary. "The Major as usual has chosen the worst camping-ground possible. If I had a dog that would lie where my bed is tonight, I would kill him, burn his collar and swear I never owned him. . . . Wet all day, and nothing dry to put on . . . I fell while trying to save my boat from a rock and have a bad cut over my left eye."

Griping, almost burned to death by a cooking fire that "took off like a prairie fire" through their pine-grove campground, scratched and tattered by afternoon cliff climbs to hunt for rock samples and plants "for the Major" and red meat for their bean-soured "innards," the party set up camp on June 28 in the delightful valley at the Green-Uinta junction. The water-slopped army flour was growing a fascinating collection of fauna. Powell, the Howlands, and Frank Goodman went off to the Uinta Indian Agency, a day's trip up the Uinta, to mail letters and bargain for food supplies.

William Byers, an editor of *The Rocky Mountain News,* was an ardent hiker and had been on the Pike's Peak climb with the Powells in 1867. The Howlands were close friends of his, and Jack Sumner was his brother-in-law. Byers and Oramel Howland, the most verbal member of the trio, had made an informal agreement—Howland would supply a running account of the expedition, which Byers would edit and publish in *The Rocky Mountain News.* Byers had correctly guessed that other news-papers would be so eager for "human-interest" stories about the race across Utah between Central Pacific and Union Pacific construction crews that little attention would be given to that "one-armed Major Somebody-or-other rowing down the Green." The ostentatious ceremony of the Golden Spike at Promontory Summit did take place on May 10, just two weeks before the Powell flotilla left Green River, and few papers bothered to mention the expedition.

But the Golden Spike story had been milked dry by late June, so telegraph operators on Chicago and Atlantic coast dailies sent copy-boys off to fetch their managing editors when their tickers began to chatter the report that "a Major Powell and all of his expedition have perished in the Colorado wilderness." The story originated in Cheyenne, where a John Risdon claimed he heard it "firsthand" from Indians who had "seen the bodies." Risdon went by train to Omaha, Chicago, and Springfield, Illinois, embellishing his tale at each stop. Neither Mrs. Powell in Detroit nor William Byers in Denver could affirm or deny Risdon's story until, in August, letters from Howland and Powell arrived from the Uinta Indian Agency. By that time, Risdon had made the Powell expedition front-page news and "Major Powell" was famous.

In the fateful letters turned over to the Indian agent, Powell told how he had bargained with squaws and at the agency store for flour, a sack of antelope jerky, and garden "sass." When the time came to start back toward the camp, Frank Goodman announced that he was quitting. He had almost drowned when the *No-Name* smashed up; the lug and tobogganing through the Canyon of the Lodore had terrified him, and he felt he could not face the dangers that might lie ahead. Powell grunted, picked up a bag of flour, and stalked off.

Through the howl of desert winds, by the towering layers of sandstone, limestone, granite, and schist, the three boats and nine men struggled through Desolation and Labyrinth canyons and on to the Orange Cliffs junction with the Colorado. Powell beached the *Emma Dean* near a grove of willows beyond the junction and ordered several days of rest in preparation for the final stage of the quest: the Grand Canyon. All of the reading he had done in the journals of conquistadors and mountain men indicated that this ride would be almost serene, since the combined power of the Green, Colorado, Little Colorado, and San Juan seemed, from chasm rims a mile above, to have carved a sinuous turquoise belt without the silver or pearl fretting of rapids and waterfalls. Thus, while the supplies were spread out and dried and the crew roused from their naps only long enough to catch fish or beaver for the cook, Powell explored the "cascade fountains and towering walls," estimated the heights and ages of the varicolored strata, collected a few fossils, and reviewed his notes and astronomical observations. The boats were reloaded on July 21, and the men were looking forward to dining on fresh provender in a Mormon settlement on the Arizona desert within a week or two.

But the river's roil through Cataract and Glen Canyons and past the Clay Hills and Waterpocket Folds of the area that would, a century later, become man-made Lake Powell was as hellish as that white inferno of the Lodore. "Row a mile of smooth water," Sumner wrote, "come to a 100 yards portage, made it to the south side, ran another mile of smooth water and got to a very bad rapid. . . . The water is about as filthy as the sewers of some large, dirty city, and stinks more than cologne ever did."

The army's salt pork eventually became rancid and maggot-infested. The Uinta Indian Agency's flour turned pea-green with mold. The last box of baking soda washed overboard. Clothing was shredded by the rocks and boot soles tore off. Yet Powell was enthralled by the Marble Canyon grottoes "all polished and fretted with strange devices" and the "baby clouds [that] creep out of side canyons, glide around points, and creep back again into more distant gorges." On August 11, Bradley wrote: "If he [Powell] can only study he will be happy without food or shelter, but the rest of us are not afflicted with it to an alarming extent."

On August 27, as they approached Grand Canyon's massive loops between the Coconino and Shivwits plateaus, they reached what Bradley described as the worst rapid yet, where "the water dashes against the left bank and then is thrown furiously back against the right. The billows are huge and I fear our boats could not ride them [even] if we could keep them off the rocks. . . . We have only subsistence for about five days. . . . There are three other [rapids] in sight below."

The next morning, after another clamber over the cliffs to examine the frothing channel beyond, the Howland brothers and Bill Dunn announced that they intended to "walk out." What they did next remains uncertain. The popular conclusion is that the Howlands and Dunn "quit the expedition in anger," but there is no evidence to support such an assumption. The Howlands were trusted friends—Oramel Howland was Powell's lieutenant and most versatile assistant in the scientific observations. It is logical to assume that Oramel Howland concluded that the expedition would be marooned in the rapids, and that he left in order to be able to lead in a rescue party of Indians or Mormon settlers.

The *Emma Dean* was abandoned. The five remaining men shot the first set of rapids in the *Maid of the Canyon* and *Kitty Clyde's Sister*, then pulled into a backwater. Bradley was standing in the prow of the *Maid of the Canyon* studying the giant wave pattern of the cataract ahead when an eddy pulled the craft back into the current, the mooring line broke, and "the loosened boat dashed out like a warhorse eager for the fray." Bradley rode the prow with the agility of a broncobuster; the *Maid of the Canyon* slap-banged through and over the spume to a deep, blue pool a quarter-mile downstream. Powell watched from the prow of *Kitty Clyde's Sister*. When Bradley stood up and waved an oar from the downstream pool, Powell cast off, crouched down and hugged the prow, and within two minutes was shaking hands with Bradley. By sundown, all of the rapids had been "bucked" and the supplies and instruments had been taken down to a grotto campsite.

Powell felt that the "race for life" was almost ended and ordered the crew to begin reloading soon after sunrise. That Sunday afternoon, August 29, "we came to a low rolling desert and saw plainly that our work of danger was done." Rowing and drifting the rest of that day and through the morning of the next, they reached the mouth of the Virgin River and the Mormon hamlet of Callville on the afternoon of August 30.

The expedition crew broke up at Callville. Bradley and Sumner returned east via Fort Yuma; the other two decided to "give *Maid of the Canyon* the full ride," and rowed the two hundred miles down to the Gulf of California. Powell and his brother rode northwest to Salt Lake City. News of "the Conquest of the Colorado" had preceded them. Congratulatory magazine offers and speaking invitations were waiting.

Correspondence, interviews, and elaboration of field notes delayed the Powells in "the bright city of Zion" for more than a week. Thus they were still there when officials of the Church of Jesus Christ of Latter-day Saints relayed a report from one of their missionaries in the Arizona Territory. The Shivwit Indians were concerned, the report said, about a tragic error made by the warriors of one of their villages on the north rim of the Grand Canyon. Drunken prospectors had recently raped and murdered one of their women. When the warriors sighted three white men crossing the desert near their village, they assumed these must be the murderers and had ambushed and killed them. Later, the real criminals were discovered. The ambushed trio had been the Howland brothers and Bill Dunn. The murders gave the Powell expedition more newspaper publicity and brought Powell even more invitations to "address our organization."

Powell spent most of the winter of 1869–70 lecturing in cities and towns from Chicago to Boston. In his talks, he introduced concepts about geography and geology that would eventually be accepted by scientists under the name of *geomorphology*. Nonscientists were speedier in their reaction to his almost reverential descriptions of those memorials to the Earth's birth spasms that walled the turbulences of the Green and the Colorado. The directors of the Atchison, Topeka & Santa Fe Railroad began to plan a spur line to the Grand Canyon's south rim. The Union Pacific's directors planned a similar line for the north rim when their "cutoff" through Salt Lake

City was completed. Congress and the Smithsonian took a second look at "that one-armed major" and decided that it would be politically expedient to give him a twelve-thousand-dollar appropriation to "complete the survey of the Colorado of the West and its tributaries."

In November, 1869, Othniel Marsh also made the front pages of the Northeast's dailies and enhanced his reputation as a pale-ontological detective. On October 16, newsmen were told, two well diggers uncovered a "fossil corpse" ten feet long on the farm of "Stub" Newell, near the village of Cardiff in upstate New York, twenty-five miles south of salt-rich Syracuse. By the time newspapermen got there, Newell had roped off the pit and put up a tent and was charging a fifty-cent fee for the privilege of gazing at this scowling, sepia-brown "giant." The story was telegraphed to New York, Boston, Philadelphia, and Washington dailies, and the name "Cardiff Giant" appeared in all the headlines. Within a week, a syndicate had been formed to dig up and display this "victim of the Great Flood," and P. T. Barnum was offering to purchase it. But it was quickly learned that Newell's brother-in-law, George Hull, held mysterious rights on both the giant and the farm and had accepted thirty thousand dollars from the syndicate for "part of his share."

The eagerness of itinerant evangelists and even pastors of city churches to label the figure as "one of the giants mentioned in the Scriptures" irked Dr. Andrew White. A native of nearby Homer and a Yale alumnus, White had toured Europe as both student and diplomat before accepting a post as professor of history at the University of Michigan. In 1867, both ability and friendship made him Ezra Cornell's choice as president of the university that Cornell was founding at Ithaca, New York, twenty miles southwest of Cardiff. Cornell University was non-sectarian; White was a champion of scientific research and, although a devout Episcopalian, was already accumulating the data that in 1896 would buttress his masterly *History of the Warfare of Science with Theology in Christendom*.

Discreet inquiries revealed that George Hull had formerly lived in Binghamton, New York, had failed there as a cigar manufacturer, and for the last year or two had been "out West." Cornell geologists reported to White that the giant, if that's what the thing really was, had fossilized into gypsum, a "most unusual" phenomenon. By this time, P. T. Barnum was displaying "The REAL Cardiff Giant" at a New York City museum—but his closest associates knew that he had had it sculptured in a week's time. The Cardiff syndicate brought their "original" to Syracuse, where railroad passengers could view it during the division stop for crew changing and refueling.

Dr. White's determination to get at the facts prompted him to persuade James Hall to come to Syracuse to examine the giant. Hull, Newell, and the syndicate were boisterously co-operative and let Hall poke and pat the oily figure as much as he wanted to. "A most marvelous thing," Hall permitted newsmen to quote. "Although not dating back to the stone age, it is nevertheless deserving of the attention of archaeologists." Hall's statement convinced James Gordon Bennett, so *The New York Herald* published a long article, quoting Hall, and pronounced the Cardiff Giant to be "genuine."

Then either Dr. White thought of that professor of paleontology over at his alma mater, or Marsh himself sensed a rare opportunity for publicity. Marsh reached Syracuse in mid-November, paid the fifty-cent admission fee, and stared at the giant. After a few minutes, he told the guard who he was and asked permission to go beyond the ropes. A syndicate official was called in and finally gave his consent; after all, a quotation from a Yale professor would look great alongside that one from Dr. Hall on the new posters.

Marsh bent over the giant, touched it with an index finger, carefully sniffed at the finger, nodded, and stepped back over the rope.

"Well?" the syndicate official asked.

Marsh scratched his beard a moment, then smiled. "Very remarkable," he purred.

"Just 'very remarkable'? Can we quote you on that?"

"No. You may quote me on this, though: a *very remarkable fake!*"

Back at his hotel, Marsh wrote Dr. White a letter and, before sealing it, let a reporter for *The Syracuse Journal* copy it. The Cardiff Giant, he wrote, "is of very recent origin, and a most decided humbug" because "gypsum is so water soluble" that even a few years in moist earth would scar it. But the surface was polished and some of the tool marks were visible, "hence the Giant must have been very recently buried."

The New York Herald reversed its stand following publication of the Marsh letter and concluded that "the testimony of Professor Marsh finally settles the claim of the monstrosity to be of antique origin." Gate receipts dropped. George Hull finally confessed that the giant had been carved in Chicago from a block of Iowa gypsum and secretly buried on Newell's farm only a year before Stub Newell began digging his well.

The Cardiff Giant exposé won Marsh the cautious respect of university and museum scholars. Their respect—and caution—increased when "Silliman's Journal" described the showmanship Marsh was using on an 1870 expedition into the West. Actually it was a grandiose elaboration of an Amos Eaton field trip. Marsh had selected twelve Yale students to join him in exploring strata and fossil beds from Nebraska to the Pacific coast. Most were prosperous enough to pay their own expenses; Yale alumni contributed funds for the few who could not. Railroad passes were begged from New York Central, Chicago & Northwestern, Union Pacific, and Central Pacific executives. A visit to the War Department in Washington produced letters to fort commanders to supply Professor Marsh and party with "adequate escort" on the searches into Nebraska, Wyoming, Utah, and Kansas back country. Each student was outfitted with a Bowie knife, a Smith & Wesson revolver, a fifty-caliber rifle, a geologist's hammer, a brass box of sulphur matches, cavalry boots and breeches, wool shirts, and broad-brimmed hats. (However, one of the twelve clung grimly to a black bowler throughout the five months.)

Soon after June graduation, the party left New Haven for Fort McPherson, near North Platte, Nebraska. Major Frank North, a detachment of his Pawnee scouts, and the most famous army scout in the West, William F. Cody, were waiting. Mounted on Indian ponies that, Cody assured them, had just been captured from a band of Cheyenne raiders, the thirteen jogged into the sand wallows toward the fossil beds on the Loup River that Hayden had detected in 1866–67. Nature embellished their introduction to the West with mosquitoes, a hailstorm, and a prairie fire. Among the ancient debris along the Loup, Marsh identified and dug out more leg and skull remnants of those "small horses"; examination under microscopes that winter indicated that he had found evidence of at least six more stages in the evolution of the Equidae.

From North Platte, the party went by rail to Cheyenne and Hayden's favorite base at Fort Russell, then crossed the Laramie Plain to Fort Bridger. At each stop the army had cavalry guards waiting, with a *remuda* of saddle-serene ponies. The ride out from Bridger was the expedition's longest because Marsh had studied enough newspaper stories about Powell's journey to suspect that huge bone yards existed on the Uinta Mountain flanks and south toward the junction of the Green and White rivers. A rich fossil bed was discovered in Bridger Basin, with evidence of such rare relics that Marsh decided it worthy of a season's dig sometime in the future. Although the struggle around the Uintas and down the Green's palisaded bleakness to the White skirted the site of Dinosaur National Monument, no bones were found; Marsh could only grumble that "there simply have to be giant skeletons in here somewhere!"

Through October and early November, the new railroad enabled comfortable surveys of the Wasatch, the ancient lake beds that had shaped the flats of northern Utah and Nevada, and that gargantuan wall that Spaniards, comparing to the massive uplift around Granada, had nostalgically named Sierra Nevada. From San Francisco, the party returned to Cheyenne, and from there took the new rail line to Denver and southeast into the Smoky Hill River country of Kansas. There Fort Wallace cavalrymen were at trackside with wagons and another *remuda* for a hunt in the Cretaceous shales up the river's north fork. The Agricultural College of Kansas had been founded in

1863; Benjamin Mudge, its professor of geology, was an astute paleontologist. In letters to Marsh, Mudge had reported evidence of mosasaur, turtle, and flora fossils in a belt of sand that jagged across the fork's eroding cliffs.

The entire party feasted on bison and antelope steaks, pork and beans, and bourbon that Thanksgiving Day, while Marsh gloated over the skulls and backbones of "great lizards, thirty or forty feet long" jutting like bleached tree stumps from the coppery blue shale.

The week before Christmas the group returned to New Haven with thirty-six bales of fossils, stories with which they would regale their home towns for months, and an eagerness to promote the "Field Trip West" as an annual event.

Marsh had an important reason for wanting to routinize field expeditions, at least for the next few years. Judging from the yield of the almost casual trip he and his students had just completed, the fossils brought in by Hayden, Powell, and King had been mere indications of the vast treasure to be found in the West; the trans-Missouri West held perhaps the richest, most fascinating store of prehistoric remains in the world. Since squabbles among architects, contractors, and museum officials threatened to delay completion of the Peabody Museum for two or three years, Marsh felt it would be worth while to train field scouts who could explore the existing fossil beds of the West and hunt out new ones. By the time the Peabody Museum opened, Marsh and his staff would have assembled the most exciting, impressive collection of paleontological wonders in America, perhaps the world.

That winter, the Smithsonian reported receipt of some beautiful fossil leaves, of a variety hitherto unknown, from the Dakota sandstones of Ellsworth County, Kansas. The discoverer and donator of these slabs was a nineteen-year-old boy named Charles H. Sternberg, a Lutheran preacher's son from New York. He had developed a passion for the mysteries hidden in rock strata during walks to and from a one-room school in the Susquehanna hills near Amos Eaton's beloved Schoharie cliffs. This was a repeat of those Lockport Gorge prowls that had

focused Marsh's life drive. There were bound to be thousands
of youngsters like Sternberg rock hunting in the high plains and
gorges in decades to come. They could be invaluable to science—*if* they were recognized and their interests properly
channeled.

The Yale field trips continued through the summers of 1871,
1872, and 1873. Their treasure yields for the Peabody included
"hollow bones like the joint of a little finger" that proved the
giant pterodactyl, or "flying dragon," had existed in North
America during the Mesozoic, and enough pieces of small and
large "dawn horses" for Marsh to conclude that Nebraska-
Wyoming-New Mexico was the region where Equus had evolved
and grown into the present-day horse.

During the same years, Ferdinand Hayden was refining his
skills in placating congressmen and delivering data to the
Smithsonian and the Department of the Interior. His budget
was increased to $25,000 for the 1870 survey of Wyoming and
adjacent territories. This enabled him to add more "specialists"
to his expedition and to respond to both scientific and bureaucratic demands for "more illustrations of the wonders of the
West."

Clarence King had scooped the field by hiring Timothy
O'Sullivan as official photographer of the 40th Parallel Survey.
O'Sullivan was one of the remarkable team of photographers
Mathew Brady had trained during the Civil War. His black-
curtained ambulance, wet plates, and enormous camera became
essential gear during the survey's treks and climbs between the
Wasatch and Sacramento. His photographs not only popularized King and the Survey but evoked sighs of admiration about
the "wonderlands" of the Pacific slope.

O'Sullivan's success encouraged Hayden to invite the landscape painter Sanford Robinson Gifford to join his 1870 march
up the Sweetwater and through South Pass. It also prompted
him to visit the Jackson brothers' shop soon after he reached
Omaha that July. The previous summer, out near Fort Bridger,
Hayden had met William H. Jackson, then on assignment as a
photographer for Union Pacific. Hayden learned that Jackson

was a New York native, born in the Adirondacks, and that he had apprenticed as a photographer's artist at Troy. The moods and details the twenty-seven-year-old was capturing along Union Pacific's high iron were exactly the kind of visual effects Hayden wanted for his survey reports. That summer and fall, Jackson took the first photographs ever published of the moody loveliness of the eagle-patrolled canyons of the Sweetwater and the agate-studded prairie known as South Pass.

Sharply southwest, down the desert trickle of the Little Sandy, Hayden surveyed, Gifford sketched, Jackson photographed, and Cyrus Thomas tallied wild game in the grama grass bleakness to the Mormons' cottonwood log ferry across the Green. On detour jogs through the sage to the Wind River foothills, Hayden located the Iron Mountain where, ninety years later, one of America's largest taconite mills would be constructed. South across the surrealistic rock and earth patterns of Red Desert, he verified the existence of the coal seams that would later bring thousands of Welsh, Cornish, and Bohemian miners to the Rock Springs region.

In Fort Bridger by September 12—they barely missed Marsh and his Yale expedition—they followed Marsh's route down Henry's Fork, flanking the Uintas, to its junction with the Green. They then backtracked up Powell's route to Green River City and returned to Washington on November 1. The collection of fossils and other specimens found en route filled sixty crates. The subsequent fourth annual report was a massive 511-page volume that included three brilliant essays on "Vertebrates of the West" by Edward Drinker Cope.

The praise evoked by Jackson's photographs, along with unabashed lobbying by the publicity staff of Jay Cooke's Northern Pacific Railroad, helped Hayden decide finally to satisfy his curiosity and devote the 1871–72 surveys to that weird region of "pewtrified trees and boiling springs" at the headwaters of the Yellowstone River. General H. D. Washburn, surveyor general of the Montana Territory, had led a party into the region during 1870, and Nathaniel P. Langford, one of its veterans, was running out of adjectives to extol its wonders

before Washington, New York, Baltimore, and Boston audiences. It should be preserved as a national park, he and his
companions felt. Trappers, army scouts, and finally the Washburn expedition determined that because the Yellowstone plateau straddled the continental divide, it must be the source of
the Yellowstone and Snake rivers, thus sluicing their alum-
tainted waters into both the Gulf of Mexico and the Pacific
Ocean.

Spencer Baird of the Smithsonian approved Hayden's pro-
posal to survey the Yellowstone area; his lobbying brought in a
forty-thousand-dollar appropriation for an 1871 expedition
from Ogden, Utah, up the Bear River's storied valley to Fort
Hall, then, skirting the snow-capped Tetons, into the Yellowstone through its back door.

Now as aware of the news value of a picture as any future
tabloid editor, Hayden not only signed on Jackson but also
persuaded the distinguished landscape painter Thomas Moran
to ride along, with all expenses paid. Hayden also added
another specialist to geological surveys by hiring Anton Schon-
born as the expedition's chief topographer. Then, taking advantage of the fat budget, he appointed Edward Drinker Cope as
principal paleontologist; this assured Cope of an outlet, via the
annual reports, for the richly detailed essays he was writing at
Haddonfield.

The seven-wagon train of the Yellowstone Survey left
Ogden's depot on June 11, 1871. Jackson's wagon was a ren-
ovated army ambulance containing a portable darkroom, five-
hundred glass-plate negatives, several cameras, tripods, and
rawhide bags filled with silver nitrate, collodion, gallic acid,
and the other chemicals essential to the "wet-plate" process.
Tethered to the wagon's tailboard was Hypo, the chunky mule
that carried the camera, plates, chemicals, tripod, and water
keg that Jackson would use on mountain climbs too arduous for
the wagon.

During the next three months Jackson and Hypo scrambled
cliffs, slithered across creeks, and teetered at the brinks of
canyons to get the best angle shots and close-ups of the Tetons,

Jackson's Hole, Devil's Den, Firehole River, the Falls of the Yellowstone, and the zany wonderland of sulphur spings, steam vents, and scalding mud jets where nature violently demonstrated some of the agonies of the continual processes of earth transformation. Thomas Moran often stalked along too, lugging pad and pencils for the sketches he would translate into memorable paintings.

Ice on the water pails, dawn temperatures of fifteen degrees above zero, and sharp earthquake tremors forced the expedition south toward Fort Hall in late August. The thousand-mile ride to and through the Yellowstone's "textbook of the past" ended at the Union Pacific station in Evanston, on the Wyoming-Utah border, in late September.

The wealth of fossils and plant specimens being unloaded from the wagons, the magnificence of Jackson's panoramic photographs, and the promise of Moran's sketches convinced Hayden that he now had the evidence needed to justify a crusade urging government protection of the Yellowstone's majesty. The conviction was transformed into a strenuous lobbying campaign when, on October 27, he received a letter from A. R. Nettleton of the Northern Pacific Railroad stating that "Judge Kelley has made a suggestion which strikes me as being an excellent one, viz. let Congress pass a bill reserving the Great Geyser Basin as a public park forever. . . . If you approve this, would such a recommendation be appropriate in your official report?"

The Smithsonian approved the idea. By mid-November, Smithsonian wives were decorating their dinner parties and afternoon tea tables with prints of Jackson photographs and inviting doubtful congressmen to "brunch" at Thomas Moran's studio to observe the progress on his epic painting, *The Falls of the Yellowstone*. Hayden outpaced Edward Cope by producing articles about "The Hot Springs and Geysers of the Yellowstone and Firehole Rivers" for *Scribner's* and *The American Journal of Science and Arts*, in addition to completing editing, writing, and make-up tasks on the fifth annual report.

The reward was momentous. In February, 1872, both the

Senate and the House passed the Yellowstone Park Bill by
unanimous votes. President Grant made it official by signing it
on March 1, thereby establishing more than two million acres of
the Yellowstone plateau as the nation's first national park, forever to be "a pleasuring ground for the benefit and enjoyment
of the people."

That same month a rascally New Yorker named Edward
Zane Carroll Judson, who since 1846 had been writing bad
novels under the name of Ned Buntline, persuaded William F.
Cody to take the leading role in his play, *The Scout of the
Plains*. He promptly gave Cody the nickname of "Buffalo Bill."
At the same time, a Cooperstown, New York, printer named
Erastus Beadle was running off large printings of paperback
thrillers called "dime novels." Buntline and Beadle became the
founding fathers of the image of the "wild West" that was
perpetuated over the next twenty-five years by thousands of
short stories and novels about gun fights, bandits, drunken cow-
boys, and wicked "Injuns"; these stories provided the formula
for the plots of countless movies, radio serials, and television
"Westerns."

But the accounts of Hayden's 1871 survey, William H. Jackson's photographs, and Thomas Moran's paintings offered a
lovelier, livelier, and far more honest heritage for the West than
the "bad guy, good guy" imagery of the pulp writers. Powell's
preoccupation with the desolate basins of the Green and Colorado rivers would advance the movement to establish national
parks at Grand Canyon, Petrified Forest, Bryce Canyon, and
Zion. Hayden's reports and Jackson's photos would be instrumental in creating the Grand Teton and Mesa Verde national
parks, just as King and O'Sullivan would influence legislation
for Yosemite, Sequoia, and Lassen national parks.

These products of geological surveys, and the weird creatures of dawn eras that Marsh, Cope, Leidy, and other paleontologists were retrieving from ancient strata, provided a Western
heritage of majestic vistas and ecological awareness worthy of
the American dream that Jefferson, Franklin, and other pioneer
dawnseekers had so exquisitely expressed a century before.

16 | Bone Barons

They were robber barons, trying to corner the old-bones market.—Nathan Reingold, *Science in Nineteenth Century America* (1964)

THE acquisitive process set in motion by the rapid mechanization of the West after 1869, and the exploitation of the West's resources, established the monopoly as an inevitable part of the American economy. But although monopolies were an expression of American values, they were also a threat to them. Theodore Roosevelt, who as a nine-year-old had contributed a squirrel skull and other treasures to the new American Museum of Natural History, was by 1908 being lauded as the "trust buster president." Some of the slyest, most ruthless industrialists of the period between 1870 and 1910 have become part of our history and its folklore under the label given to them in 1934 by Matthew Josephson: the robber barons.

Again, paleontology was in the vanguard. The profession experienced both "trust" efforts and "barons" between 1872 and 1897. The ensuing struggle brought trainloads of fossils to museums and added vastly to man's knowledge about the Earth's age and processes of evolution. The barons of this struggle were Edward Drinker Cope and Othniel Marsh.

The twenty-five-year feud between the Quaker genius and the Yale professor matured in 1871, while Hayden and Powell were exploring their Yellowstone and Grand Canyon "wonderlands." Cope's appointment as official paleontologist of the Hayden surveys did not necessitate field trips. But eagerness to see the landscapes that were yielding all of those amazing fossils, from Cambrian trilobites to Pleistocene tigers and camels, lured Cope out to Kansas during the summer of 1871. He belligerently chose the chalk beds of the Smoky River for his research, knowing that Marsh and the Yale group had already searched there. Although he failed to detect any evidence of pterodactyl, the "flying dragon" that was temporarily exciting Marsh more than the dawn-horse bones, he did fill two large wagons with skeletons of sea turtles, the eight-hundred-pound "bulldog tarpon," *Portheus,* and other extinct species. That fall he openly bragged, "I secured a large proportion of the extinct vertebrate species of Kansas, although Professor Marsh had been there previously."

Joseph Leidy decided that he would finally yield to his urge to dig in the West and arranged his schedule so that in 1872 he could join Mr. and Mrs. Cope and their six-year-old daughter, Julia, on a summer-long expedition. Hayden had found evidence of fossil remains in the Eocene badlands of Bridger Basin, near Fort Bridger, and had brought back samples of the petrified fish and other marine life that were sandwiched into the shales of Union Pacific's Petrified Fish Cut. But Leidy knew that Marsh and his party had explored there too; Cope's antagonism toward Marsh worried him. Thus, in March, 1872, letters were exchanged between Marsh, Cope, and Leidy in which each promised to mail to the others copies of "any papers on the subject [of fossils from the Wyoming Eocene strata] we might issue, the date of publication to be written, or printed, on each pamphlet."

The Cope-Leidy and Marsh teams reached Fort Bridger a few days apart, shaped up their trains, and rattled off to find "dig" sites. Marsh was serene about this "invasion of our territory" both because of the written agreement about publication of data and because at least one of the soldiers assigned to escort the Cope-Leidy train was on his (Marsh's) payroll. The truce ended in late July, after Joseph Leidy found a fossil of an Eocene beast that was almost as large as an elephant and was equipped with three horns as well as tusks. Leidy wrote to Philadelphia about the discovery and dutifully sent a copy of his letter to the Marsh camp.

Marsh wrote a letter to "Silliman's Journal" challenging some of the terminology Leidy had used. He also ordered his entire crew to concentrate on finding remains of the strange beast. But Cope soon dug out an almost perfect skull of the species, telegraphed news of his discovery to a Philadelphia newspaper, and sent analytic descriptions of the probable skeleton to both the American Philosophical Society and "Silliman's Journal." And he blithely neglected to notify Marsh.

Marsh had taken the precaution of generously tipping the telegrapher some time before, so he received a copy of Cope's wire. Marsh sent a telegram east pointing out errors in Cope's identification of the find and in his spelling of the beast's Latin name.

Thereafter, the battle became increasingly intense. Cope spent hours each day on a hilltop spying on the Marsh dig. This encouraged Marsh's crew to assemble a skull from the jawbones, teeth, eye sockets, and horns of a dozen species. They buried "Old What-you-may-call-it" just before Cope showed up for his daily spell at the telescope. When he did arrive, they put on an elaborate pantomime of arduous shoveling and great excitement. Cope sneaked over that dusk, dug up "What-you-may-call-it" and wrote a paper about its "significance."

Joseph Leidy soon found it "urgent" that he return to Philadelphia. For the remaining two decades of his life, Leidy assiduously avoided involvement in the bone feud and concentrated on completion of his *Contributions to the Extinct Vertebrate Fauna of the Western Territories, Freshwater Rhizopods of North America,* and other definitive works.

During the winter of 1871–73, Marsh and Cope each published sixteen articles about the skeleton with horns as well as tusks that Leidy had discovered. They gave different names to the beast, which was eventually accepted as *Uintatherium,* a large, rhinolike creature that became extinct at the close of the Eocene.

In its June, 1873, issue, *The American Naturalist* published Marsh's accusation that "Prof. Cope's errors will continue to invite correction, but these like his blunders are hydra-headed, and life is really too short to spend valuable time in such an ungracious task, especially as in the present case Prof. Cope has not even returned thanks for the correction of nearly half a hundred errors." Cope, in retort, alluded to "the learned professor of Copeology at Yale."

In retrospect, it seems plausible that both the wealth and the public awareness of American paleontology might have been held back for decades if Cope and Marsh had not engaged in their theatrical, and often farcical, bone battles precisely when they did. Basically, and for the eventual good of the cause they shared, they complemented each other. Cope was an intense genius of vivid imagination with a tendency to rush off on intellectual tangents. Although a devout Quaker, he was quick-tempered and often combative. Marsh, on the other hand, was a pedant, glumly methodical about assembling details, rudely belligerent about scientific errors, and as deft as a Tammany Hall boss in political maneuvering and lobbying. Also, he was a bachelor who seemed to be remarkably free of the need for family life.

When the Cope-Marsh arguments grew so strident that they became a source of tongue-in-cheek human-interest copy for newspapers, each principal realized that he had achieved enough status as a public figure to be able, literally, to "make news." Consequently, each engaged in spectacular "stunts" and allegations that kept their names in the headlines and on the editorial pages. These acts, naturally, also popularized the dawn creatures they were zealously crating east to Haddonfield and New Haven.

Confident of his own prowess as a lobbyist and publicist, Ferdinand Hayden made no attempt to shy away from the feud, so Cope was retained as official paleontologist of Hayden's survey. But Cope never accompanied Hayden on the treks through the Rockies. Instead, using Hayden's stratigraphic maps as guides, he organized—and largely paid for—his own expeditions to Wyoming, Colorado, and that remote Judith River area where Hayden had found the first American dinosaur teeth.

The only experience Cope had with a U.S. Geological Survey team came in 1874, when he left Hayden's staff to accompany Lieutenant G. M. Wheeler's survey west of the 100th Meridian.

The Wheeler survey was politically suspect—and still is. Before the Civil War, government-sponsored surveys in the West had been organized and administered by the War Department. This was true of Frémont's "pathfinder" expeditions and of the Railroad Route Surveys administered by Secretary of War Jefferson Davis. The data obtained were considered "essential to the national defense," and hence were top secret. But the lobbying of Spencer Baird and "Black Jack" Logan transferred federal surveys to the Department of the Interior and transformed them into scientific teams specializing in such nonmilitary matters as geological data, soil analyses, location of mineral veins, and old-bone collecting.

Early in 1871, the army's Corps of Engineers announced the creation of a survey to obtain "correct topographical knowledge . . . and prepare accurate maps . . . of those portions of the United States Territory lying south of the Central Pacific Railroad, embracing parts of Eastern Nevada and Arizona." The "Big Four" who controlled Central Pacific were already surveying a trans-Arizona line toward New Orleans; the Arizona highlands were the new redoubt of the Apaches, the fiercest Indian aggregation in the West. The survey would be commanded, the secretary of war announced, by Lieutenant George M. Wheeler, currently chief engineer for the army's Department of California.

The order for an army survey in the West was met with anxiety in scientific journals. Was this a move toward returning *all* of the federal surveys to army jurisdiction? What would happen to the botanical, zoological, paleontological, and geological data coming out of the Hayden and King reports and soon to come from the Powell survey of the Colorado Basin? Why didn't these War Department plotters let professional scientists handle scientific matters?

Wheeler's orders were remarkably like the orders given to Hayden, King, and Powell. The geological formations were to be explored and the potentials for mining and agriculture detailed. The expedition that assembled in the bleak mining town of Elko, Nevada, in the spring of 1871 included Grove Karl

Gilbert—soon to become Powell's most trusted associate—and
Archibald Marvine. Both were excellent geologists. Lieutenant
Wheeler had outbid Clarence King for the magnificent skills of
Timothy O'Sullivan, the photographer, and had also signed on
a Boston Back Bay aristocrat named Frederick W. Loring to
report the survey's achievements to newspapers and scientific
journals.

But hearsay reports published in Pacific slope newspapers in
1872 and 1873 accused Wheeler of torturing Indians, hanging
a boy herdsman by his thumbs to "learn the whereabouts of a
lost mule," and imposing martinet discipline on personnel.
"The Wheeler Exploring Party seems to have been composed of
brutes, if not worse," one editor fumed. "[It] has disgraced the
U.S. service."

With these allegations on record, the mystery of Cope's
acceptance of the post as paleontologist of the survey remains
unexplained. Some historians have contended that he had used
up most of his funds financing his 1872–73 expeditions and
needed the salary Wheeler offered. The best of all paleontologic
historians, George Gaylord Simpson, offered a more plausible
reason by pointing out that "one of Wheeler's geologists found
traces of a new mammalian fauna in the Santa Fe beds"; therefore, continued Simpson, "is it not likely that the lure of fossils
in New Mexico . . . motivated Cope's temporary transfer
from Hayden to Wheeler?" Cope's persistent defiance of
Wheeler's orders and his discovery of "the most important find
in geology I ever made" in beds even older than the Santa Fe
(Miocene) marls, substantiate Dr. Simpson's hypothesis. It
may be assumed, then, that Cope was intent on locating an
Eocene bone yard as yet unknown to "that Yale professor" or
his spies.

When Cope reached Pueblo, Colorado, on July 25, 1874, he
learned that his group would be commanded by a zoologist
named Yarrow and would consist of five men, including a
teamster and a cook. Lieutenant Wheeler's orders emphasized
that Cope would serve only as the geologist of this topographical exploration southwest through the Sangre de Cristo Moun-

tains toward Sante Fe. Cope held his tongue, managing to act the meek introvert in Yarrow's presence, then loosed the invectives in letters to his father and his wife.

On August 10, at Rito, Cope found the tooth of an Eocene mammal. He called the beast *Bathmodon,* but it was later given the name of *Coryphodon,* because of British priority in discovering the genus. At Rito he met Father Antonio Lamy, "a very intelligent priest," who showed him his collection of fossil bones and teeth, also of *Coryphodon.* Cope was elated. This extended the lowest Eocene "500 miles south of where known!" He asked the priest about where the specimens had been found —and his answers wove visions of a major new fossil field in the offing.

There followed a quick trip into the Pojoaque badlands, not far from modern Los Alamos, which netted some important Miocene vertebrates. Then, on August 15, Cope skirted the Pueblo cornfields to Santa Fe and requested an audience with the Gettysburg hero David M. Gregg, then commandant of the army's Department of Santa Fe. He so convinced General Gregg of "the vital importance to science" of the Eocene bone yard that the general countermanded Lieutenant Wheeler's orders and assured Cope's freedom to explore the area thoroughly.

Yarrow did not learn of Cope's audacity until September 1; it influenced his "transfer" back to Washington two weeks later. Cope and Wheeler finally met on September 14. Wheeler had already been briefed by General Gregg, and, lamenting his inability to imprison a civilian "deserter," sent Cope back to the fossil hunt.

The arroyos south of Yegua Canyon, near present-day Lindrith and the Jicarilla Reservation, provided the richest deposits. "As soon as we picketed horses, we began to find fossil bones," Cope wrote his father. "By sundown I had twenty spec. of Vertebrates! all of the lowest Eocene, lower than the lowest of Fort Bridger. The most important find in geology I have ever made, and the paleontology promises grandly."

The search continued through the early November freeze-up,

its intensity relieved by days when Cope examined abandoned Indian villages, made pen sketches of the countryside, and composed reports of his discoveries to be posted over the mountains to Lieutenant Wheeler's camp. Cope and the bales of fossils did not reach Philadelphia until Thanksgiving week. The fossils were accepted, even by Marsh, as "classical types for the extremely important early Eocene faunas." The first *Report upon Vertebrate Fossils Discovered in New Mexico*, which had been sent to Wheeler in late September, was hurriedly approved and was in print before Cope reached Haddonfield.

But subsequent data irked Wheeler, and he began writing Cope waspish letters, which stimulated Cope to even more mischievous retorts. The "Cope burden" that disrupted the routines of Wheeler's editing chores that winter grew out of Cope's insistence on inserting snide comments about Othniel Marsh into his copy. After crossing out a dozen of these jibes, Wheeler sent Cope a letter declaring: "I cannot permit that your report or that of any person received . . . shall by the slightest inference reflect upon anyone engaged in similar work. If you wish to reflect upon Prof. Marsh you must do it outside of reports published by this office."

A few weeks later, Wheeler's teeth gritted again as he composed an answer to a complaint from Professor Marsh that the Wheeler survey was being unfair "to this institution" by failing to send adequate samples of its paleontological discoveries.

In 1876, Marsh hired David Baldwin, a talented paleontologist, and sent him out to Cope's sites in the San Juan basin. Baldwin shipped so many mammal remains to Yale between 1877 and 1880 that Cope was moved to outbid Marsh for his services. The mammal remains that Baldwin routed to Cope between 1881 and 1888 enabled Cope, as Simpson pointed out in a speech to Philadelphia's Academy of Natural Sciences in 1951, to add "a whole new epoch, the Paleocene, to the history of mammals."

Naturally, an invitation to rejoin the field forces of Wheeler's survey in 1875 was not forthcoming. Cope concentrated on plans and notes for the series of books he planned to do for the Hayden survey reports. He had already decided on *The Vertebrata of the Tertiary Formations of the West* as, at least, a worthy working title for the first volume. The salary he received from the Wheeler survey had barely paid his expenses for the 1874 journey, and his pride would not let him appeal to his father for still another loan. He began to write articles about the mysteries of prehistoric life for nonscientific magazines and newspapers. But the financial crisis was not desperate enough for him to seek university employment. The Haverford experience still distressed him. So did that one unfortunate day when he made discreet inquiries about teaching at Princeton—still officially called the College of New Jersey—only to be bluntly told that his "radical views" were not acceptable to the "New Light" Presbyterians on its board of trustees.

The financial distress ended that summer with the death of Cope's father. Cope's share of the estate included investments worth more than $250,000. In view of the titanic rivalry that would arise in Colorado and Wyoming in 1877, fate was being awesome beyond coincidence.

The relationship between Alfred Cope and his son Edward had been so rich and multifaceted that it would be rare in any age. They understood, respected, and loved each other. It is probable that Edward Cope's grief was so deep and persistent, and his sense of the disinterestedness of science so great, that he felt no desire to engage in the widely publicized struggle between Othniel Marsh and the Interior Department's Bureau of Indian Affairs.

It is suggestive of Marsh's disdain for obstacles that he had chosen November, 1874, as the month to begin a fossil hunt in Ferdinand Hayden's beloved Dakota Badlands. Incipient freeze or not, Marsh would move. That summer, Colonel George Armstrong Custer had led an expeditionary force of more than a thousand cavalry into the Black Hills and had permitted a natural scientist or two to ride along. The scientists found evidence of gold quartz in some of the gulches; this electrifying news reached newspaper correspondents. At the same time, the Northern Pacific Railroad was pushing its main line into Dakota lands that had been solemnly pledged, in an 1868 treaty, to be "the redman's forever."

Custer was detested by most of the Sioux for the atrocities he permitted at the Battle of Washita in 1868 and for subsequent brutality. The imminence of a gold rush into the Black Hills and the rape of their land by the Northern Pacific loosed angry demands to "take to the warpath." Red Cloud, the fifty-two-year-old senior chief of the Oglala Sioux, had organized the 1866–68 campaign that caused abandonment of the forts guarding the Bozeman Trail into Montana and led to the negotiations for the 1868 treaty. By November, 1874, Red Cloud was pleading with Sitting Bull, Crazy Horse, Gall, and other advocates of all-out war to reconsider and make careful plans. These furtive conferences, held in wintertime shacks a day's ride from the Black Hills, were interrupted by the rattle of Marsh's wagon train and by Marsh's request to search for the "thunder horse" along the White River.

Convinced that this bumptious fat man must be the first of the gold rush despoilers, some of the Sioux became so belligerent that the white resident agent ordered Marsh and his escort to return to railhead. Marsh refused, sought out Chief Red Cloud, and, through an interpreter, proposed a "great feast" and a discussion of his expedition's purpose. The fat man's haughty disobedience of the agent's orders intrigued Red Cloud. He accepted.

Marsh's conduct during the feast and his after-dinner speech won the Sioux's grudging consent to the Badlands dig, although no warrior escort was provided. Well guided by notes jotted from Hayden's writings and by Red Cloud's advice, Marsh was able to pick up wagonloads of fossils. He deliberately chose a return route that took him past the Indian Agency so that Red Cloud and his councilmen could see that no samples of gold-bearing quartz were being smuggled off to assayers.

Red Cloud was so impressed by this semihonest white man that he invited Marsh to take a tour of the Sioux huts and examine the flour, pork, cloth, and other subsistence supplies provided by the Bureau of Indian Affairs. The stench of the putrefying pork, rat dung in the flour, and dry rot of the cloth, plus councilmen's testimony about the cheating and double-dealings of government agents and contractors sent Marsh east in a rage against "our vile bureaucrats."

Early in the spring of 1875, Red Cloud got a message through to Marsh reporting that no coffee, sugar, or flour rations had been distributed by the agent since Marsh left, and that some Sioux families were actually starving, while others were suffering from rickets and other evidence of malnutrition.

Marsh promptly took the train to New York City, where he found an eager listener in James Gordon Bennett of *The New York Herald*. As firm as he had been almost a decade before, when he had ordered Henry Stanley to go and find Livingstone, Bennett assigned reporters to investigate "the Indian ring" at the Department of the Interior, instructed his editors to give Marsh full co-operation, and sent one of his best interviewers off with Marsh "to pump every speck of fact you can out of the professor."

Between April and September, the *Herald* published more than forty articles about the "far-flung corruption" that had become "a constant source of discontent and hostility among the Indians." Other newspapers and magazines joined the crusade. Charges and countercharges grew vituperous; the Interior Department's publicity men "leaked" the allegation that Marsh was a "wanton bachelor" and had been guilty of "certain unmentionable indiscretions during his Western trip."

But early in September, President Grant called in Christopher Delano, secretary of the interior. He accepted "regretfully" Delano's resignation and growled orders for "a house cleaning of that whole Indian Bureau."

Both the *Herald* and the dignified *Boston Transcript* extolled Marsh for "a great public service, and *The Nation* editorialized: "Considering . . . there was only on the side of the Indians and the people of the United States a college professor doing work to which he had never been trained, and incited only by the desire to do his duty, and having against him the united forces of all the corruption, all the ignorance and all the prejudice of the 'practical men' of the Plains . . . the results are startling."

The best reward, however, came in the form of a carefully carved and polished peace pipe and a translated message that hailed Marsh as "the best white man I ever saw." It was signed with Red Cloud's symbol. In 1880, Red Cloud came to Yale, was Marsh's house guest for several days, and posed in full regalia for a painting that had a permanent place of honor in Marsh's study.

His Quaker convictions prompted Edward Cope to give his grudging approval as he followed the newspaper accounts of his archrival's championship of the Sioux. But the articles worried him, too. Marsh's choice of the Dakota Badlands for that November dig indicated that he might also be planning intensive searches of the formations that Hayden had pioneered on the Judith and upper Missouri of Montana. This area, too, was claimed by the Sioux, as well as by the Blackfoot and Crow Indians. The formations were of special interest to Cope because Hayden had discovered dinosaur teeth there, whereas Cope had failed to find any in the Eocene strata of the San Juan basin. If the Montana expanses did contain Cretaceous rocks, this finding would indicate that they were the same age as those Haddonfield marls where Cope had picked up so many dinosaur bones and teeth. The Cretaceous was presumed to be millions of years older than the Eocene. Did dinosaurs become extinct at the end of the Cretaceous? Did they evolve into different types of reptiles?

This line of thought repeatedly brought Cope back to the letter he had received in May, 1876, from young Charles Sternberg. The fossil leaves Sternberg had sent to the Smithsonian in 1870, the letter explained, were merely a first step toward his decision "that whatever it cost me in privation, danger and solitude, I would make it my business to collect facts from the crust of the earth." In 1875, Sternberg had studied paleontology under one of Marsh's scouts, Professor Benjamin Mudge, at Kansas State Agricultural College. In the spring of 1876, Mudge rejected Sternberg's request to join a dig that Marsh was financing in western Kansas. Thus, "in a terrible state of suspense," Sternberg wrote Cope asking for an advance of three hundred dollars to buy a rig, hire a cook and driver, and devote the summer to digging marine fossils for Cope in the Cretaceous strata of the Flint Hills in southeastern Kansas. Alfred Cope's will had by this time been probated; the check was sent to Sternberg.

Cope reached a decision in early July. He telegraphed Sternberg and his associate, J. C. Isaac, asking them to meet him at the Omaha depot prepared for a two- or three-month expedition in the North.

In Philadelphia, hundreds of people were hastening to newspaper offices each morning to read the latest bulletins about the Montana disaster. On June 25, the War Department finally admitted, Sitting Bull, Crazy Horse, Gall, and three thousand Sioux and Cheyenne warriors had ambushed Colonel Custer and six hundred veterans of the Seventh Cavalry in the valley of the Little Big Horn River, northeast of the new Yellowstone National Park. The brutalities ascribed to "Yellowhair" were finally avenged; Custer and more than two hundred of the Seventh's finest had died on a Little Big Horn hillside.

The headwaters of the Judith were only 150 miles northwest of the battlefield. But Cope did not hesitate. If Marsh could negotiate with the Sioux, Cope could. Sternberg, Isaac, and he would search the Judith exposures for dinosaur remains until winter forced them out.

Railroads, stagecoaches, and wagons took a week to make the trip from Omaha to Fort Benton, an army post forty miles east of the Missouri's Great Falls and eighty miles west of the Judith-Missouri junction. Fort Benton's officers were too busy with defense plans to give Cope more than a brusque warning. Sitting Bull was reported to be leading a large party of Sioux northward in an effort to reach Canada; troops from Forts Ellis and Abraham Lincoln were trying to cut them off. Meanwhile, the Crows were gathering in the Judith's basin for the berry and nut harvest and their fall hunt; there might be more than a thousand of them.

Cope thanked the officers for the advice and waited a day for an order forbidding him to head for the Judith. It didn't come, so he bought 'worn out mustangs and a balky colt," a second-hand box wagon and a tent, and he hired a cook and a half-breed scout. The outfit sloughed through thunderstorms for four and a half days to reach the Cretaceous bluffs at the mouth of the Judith. They found a tepee village of more than two thousand Crows pitched on the riverside meadow nearby.

A Crow youth was hanging about the next morning as Cope shaved and bathed before breakfast. When Cope took out his false teeth and began scrubbing them, the boy rushed off to the Crow encampment. A few moments later, a party of warriors galloped up to the wagon, gave the outstretched-hand signal of peace, dismounted, and stalked toward the paleontologist. "Please," asked one of the warriors, "do that trick with your teeth again." Each morning thereafter, scores of Crows gathered politely to observe "Magic Tooth" at his ablutions. There was no thievery or jeering at the Cope camp, but there were occasional gifts of sage hens or fresh fish for Magic Tooth.

Slowly working the outcrops, Cope, Sternberg, and Isaac pointed themselves into the badland country east of the Judith. They found teeth and skeletal parts of dinosaurs far larger than the ones from the Haddonfield peat. One of the skulls had horns. Another kind of dinosaur had a rounded, oblong jaw, shaped like a duck's.

In the erosional maze of Dog Creek, Sternberg instituted a packaging technique that would become essential to all paleontological expeditions. Epochal seepage and submersions had mineralized most of the dinosaur bones into agate, jasper, and other brittle compounds. Picks, hammers, crowbars, even whisk brooms were used to determine the extent of the skeletal parts and to free them from the rocks. Then they had to be transported by wagon to railroad sidings or boat landings and jolted two thousand miles to Philadelphia. The amount of breakage was ghastly. One afternoon, as Sternberg stood beside the wagon pondering the hazards of all the digging and shipping, he realized that he was leaning against a burlap sack filled with rice. This staple of the Oriental diet, still the most portable carbohydrate, was an essential item in the chuck wagon of any trek into the Far West. Neither Cope nor Sternberg nor Isaac particularly liked rice, so the bags still lay in the wagon, taking up room that could be used for fossils. Rice, Sternberg now recalled, could be boiled into a gelatinous goo that was a moderately good substitute for glue.

Sternberg reported his idea to Cope that evening. Cope walked over to the wagon, stared at the rice sacks, and ordered the cook to start boiling a kettle of water. Before bedtime, Sternberg and Cope and the puzzled cook had the horned dinosaur skull swathed in burlap strips that had been soaked in rice paste. By morning, the mess had hardened. The skull, looking like an Egyptian mummy, was laid at the front end of the wagon bed and served as the base for other rice-and-burlap-swathed bones. Improvements on this technique in 1877 and 1878 merely substituted flour paste or plaster of Paris for the rice glue. The breakage ratio for fossils dropped remarkably.

By late September, almost a ton of dinosaur bones and other Cretaceous relics had been collected and encased in rice-paste bandages. The schedules of Missouri River steamers were as uncertain as the weather. Cope decided it was time to attempt the forty-mile haul over the bluffs to Cow Island, where steamers usually anchored overnight. But he returned to camp one sundown to find that the half-breed guide and the cook had packed their gear. A party of Crows had ridden into the camp an hour before, the guide muttered as he stuffed supplies into his saddlebags, to warn Magic Tooth that Sitting Bull and his Sioux had entered the area moving north, probably to cross the Missouri at or near Cow Island. The professor and Sternberg and Isaac could ride along if they wanted to—but not with that wagonful of old bones.

Cope nodded, walked back to the tent, and reappeared a moment later with a few greenbacks and silver pieces in each hand. He gave one handful to the guide and the second to the cook and hurried back to the wagon's tailboard. He remained there, squinting and exclaiming over the day's harvest of fossils, as the guide and cook clattered off.

The search continued for another week, with Cope and Sternberg alternately doing the cooking and clean-up chores. Isaac patrolled with rifle cocked during his night watches. So did Sternberg. But when it was Cope's turn to do guard duty, he sat up, asked about the weather, and went back to sleep.

Only "guts" and Cope's inexhaustible patience got the wagon across the ravines and down the cliffs to the Missouri shore. Cope was waiting at the Cow Island landing when the eastbound S.S. *Josephine* splashed in. He went aboard and offered the captain an extra fee if the *Josephine* could be anchored a few miles downstream while the wagonload of fossils was ferried aboard. The captain refused.

Cope found a leaky scow beached near the wharf and paid an exorbitant price for it. He and Sternberg floated and poled it down river, beached it, and loaded it with the fossils. Then, as keelboatmen on the Missouri and Mississippi had been doing for almost a century, they hitched one of the horses on by a long rope and cordelled the scow upstream to the Cow Island wharf.

"When about sundown," Sternberg recalled in his autobiography, *The Life of a Fossil Hunter,* "we hove-to under the big steamer, the deck was crowded with passengers watching our approach. Cope was covered with mud from head to foot, and his clothing, with hardly a seam whole, hung from him in wet, dirty rags. He had forgotten to bring along any winter wearing apparel so, although the nights were quite cold and the women were clad in fur coats and the men in ulsters, he emerged from the sergeant's tent, whither he had carried his grip, in a summer suit and linen duster."

Sternberg and Isaac returned to the wagon, got it back up the cliffs, and continued the fossil search until snows forced them into Fort Benton. Sitting Bull and his Custer-avengers crossed into Canada the following February and stayed there five years.

Cope's determination to find dinosaurs in the Montana Cretaceous was prophetic; 1877 was the Year of the Dinosaurs. The extraordinary sequence of discoveries began in the Fosse Sainte-Barbe coal mine at Bernissart, Belgium, near the French border. A new gallery being dug at the 1,046-foot level revealed so many giant bones that the Royal Museum of Natural History was called in to analyze them. Maurice De Pauw and a team of osteologists deduced that the miners were approaching what had been a clay gully during the Cretaceous period. Scores of iguanodonts seemed to have mired and died there. It took four years of tunneling to remove these remarkably preserved and complete skeletons. One of the rewards was an exhibit of eleven standing skeletons, still on display at the Royal Museum in Brussels.

The owners of the Fosse Sainte-Barbe were still wondering whether they should bother notifying Brussels of their discovery when both Marsh and Cope received boxes of bones from Morrison, Colorado. The shipper was a schoolteacher named Arthur Lakes, an Oxford University alumnus who had joined the rush of British cattlemen and thrill seekers to the trans-Missouri West. Lakes, who was fascinated by fossils, had discovered an outcropping of giant bones in one of the foothills near Morrison. Both Cope and Marsh recognized the pieces Lakes had shipped to them as dinosaur bones; each reacted typically. Marsh sent Professor Mudge out from Kansas to persuade Lakes to join Marsh's staff and continue his hunt exclusively for the Peabody Museum; Cope analyzed his box of bones and wrote an article about them for the American Philosophical Society.

The article was still on press when Cope received a box of dinosaur bones from O. W. Lucas, superintendent of schools in Canyon City, Colorado. Lucas had found the bones in the Royal Gorge area of the Arkansas River; they were larger and more numerous than Lakes's find. Lucas hired a crew and began quarrying at the site, collecting exclusively for Cope. By August, Cope had enough samples to conclude that the Canyon City dinosaurs had been vegetarians, sixty to seventy feet long, exceeding "in proportions any other land animal hitherto described, including the one found near Golden City by Professor Lakes."

When news of the Canyon City dig reached New Haven, Marsh instructed Professor Mudge to go to Canyon City immediately, determine the stratum in which Lucas and his crew were digging, locate another outcropping of it, and "dig for Yale."

Mudge's assistant at the Canyon City dig, begun in September, was Samuel Wendell Williston, a "first-generation" Kansan. He was born at Roxbury, Massachusetts, and had been brought to Kansas as a five-year-old. A casual interest in the fossil shells embedded in the ridges near his Manhattan home grew into a passion that determined his life goal and led to his decision to study geology under Professor Mudge at Kansas State Agricultural College. A reading of Sir Charles Lyell's *Antiquity of Man* incited a turmoil of doubt and guilt within him because of the Baptist convictions of his parents. Attendance at a lecture supporting Darwinism given in the local Congregational church (Williston later contended it was "the first lecture in favor of Evolution ever given west of the Mississippi") persuaded him to read *The Descent of Man* as well as some writings by Huxley and Spencer. In spite of Professor Mudge's great disdain for Darwin, Williston was allowed to become Mudge's assistant on the field trips for Marsh and, in March, 1876, signed a three-year contract to work exclusively for Marsh. His salary was listed as forty dollars a month.

The discovery of "the hind leg, pelvis and much of the tail" of a skeleton eighty feet long was so exciting to Marsh and Mudge that Williston was assigned to take "the innumerable pieces of bone" to Marsh at Yale. "I did observe that the caudal vertebrae had very peculiar chevrons, unlike others I had seen," Williston recalled, "and so I attempted to save some samples of them by pasting them up with thick layers of paper. Had we only known of plaster-of-paris and burlap, the whole specimen might easily have been saved. When I reached New Haven, I took off the paper and called Professor Marsh's attention to the strange chevrons." Marsh pondered the oddity, asked Williston about details of the skeleton, and decided that this was still another family of "terrible lizard." He named it *Diplodocus*.

Williston's acuity in spotting the unusual chevron bones of *Diplodocus* impressed Marsh. A few weeks later, he sent Williston a wire asking him to leave immediately for Como, Wyoming. He was to find two men named Harlow and Edwards and "collect and learn all possible" about the dinosaur bones they claimed to have. Thus, at twenty-five, Samuel Wendell Williston blazed the trail to a dinosaur graveyard that extended for seven miles and contained tons of bones.

Como was a small station-and-cabin town on the Laramie Plain that, more than a mile above sea level, ripples between the gaunt Laramie and Medicine Bow ranges. Nine years before, Hell on Wheels had curved Union Pacific's high iron almost due north up the plain to reach the Red Desert crossing of the continental divide into the Green River valley. Como thus became a section house, station, and siding for the hauls between Laramie and Medicine Bow. The tan bluffs glooming to the north, Ferdinand Hayden and others had concluded, had been formed during the Jurassic period of the Mesozoic era. Eventually the geologists decided that Como's cliffs were 140,000,000 years old, and were the same strata as the dig sites at Canyon City.

Williston had little difficulty in locating Harlow and Edwards. A bin of dinosaur bones sat in a corner of the station's freight office. After a three-minute argument, Williston showed the station agent the telegram from Marsh. The agent grinned, offered a handshake, and announced that he and the section foreman were Harlow and Edwards. The reason for the pseudonyms was never fully explained but seems to have been associated with guilt feelings occasioned by "a bit of antelope hunting" during working hours. The station agent was William E. Carlin. The section foreman was spry, slightly bowlegged William H. Reed, whose huge black mustache framed two-thirds of his nose.

Reed guided Williston up a bluff near the station to an outcropping of leg bones, vertebrae, and skull bones that were
"very thick, well preserved and easy to get out." A few days
later, Williston wrote Marsh that "there are five hundred
square miles of fossil country. . . . They have found bones
fifteen, eighteen and more miles from here in different directions. Friday I found an entire forearm lying exposed above the
ground seven miles east from here & literally acres of clay wash
covered with fragments. Whenever a surface is exposed, there
are fragments."

Marsh paid Carlin's expenses to New Haven so that an agreement could be worked out. He offered ninety dollars a month
each to Carlin and Reed if they would dig out bones and bring
them almost a mile to a railroad siding, where they would be
loaded onto a handcar. The contract was for a year; there were
two important stipulations. First, Carlin and Reed must main-
tain a strict watch on the dig area and keep out other collectors.
Second, Marsh must have a personal representative at the dig as
his superintendent. The contract was signed. Williston became
the forty-dollar-a-month superintendent of two ninety-dollar-a-
month diggers.

The bones that Carlin and Reed had already stacked in the
Como freight office were of so many different varieties that
Marsh was able to rush a spectacular article into the December
issue of "Silliman's Journal." It identified four new dinosaurs:
"*Stegosaurus*, most bizarre of land animals, with small head,
body heavily armed with large bony plates, two rows of huge
plates standing erect on its back, and tail bristling with huge
spikes—an impressive defensive armament for a harmless
plant-eater; *Apatosaurus*, a dinosaur between fifty and sixty
feet long; *Allosaurus*, a fierce carnivore, half the size of the
foregoing, but much more terrible in claw and tooth; and in
striking contrast, the graceful little *Nanosaurus* and the tiny
leaping *Hallopus*."

The Marsh quarrying at Como Bluff would continue for
more than a decade and, like the Cope and Marsh quarryings at
Canyon City, would serve two historic purposes in the develop-

ment and popularization of paleontology. From them, as from the tunnels being cautiously drilled in the depths of the Fosse Sainte-Barbe mine, would come the first *complete* skeletons of the vegetarian and carnivorous giants that dominated the Earth's land life for more than 100,000,000 years. Equally important, these digs became the proving grounds for young-sters, like Charles Sternberg and Samuel Williston, who would expand the search for paleontological evidence of the Earth's age and dawn eras from the Arctic to Patagonia, from Africa to the Mongolian deserts.

By the winter of 1877, the focus of paleontology's future was moving east to the laboratories at Peabody, the University of Pennsylvania, and to Cope's crate-stacked workrooms at Had-donfield. The rivalry between the two bone barons would lead to political vendettas that would bring shame and sorrow to each of them. Still, throughout the remaining two decades of their feud, their dedication to their goals never abated as Marsh and Cope pioneered the intricate skills of reconstruction.

17 | The Resurrectionists

The hunter of wild game is always bringing live animals nearer to death and extinction, whereas the fossil hunter is always seeking to bring extinct animals to life.—Henry Fairfield Osborn

CASPAR WISTAR and Charles Willson Peale needed six months—from November, 1801, through April, 1802—to puzzle out and assemble the two great mastodon skeletons from the jumble of bones, tusks, and teeth carted out of the Wallkill valley. Even then, they were so uncertain of the texture of the beast's hide that only the skeletons, with a few hand-carved ribs and bones, went on public display. Yet their task was simplified by the evidence that the beast was obviously related to the elephant, so drawings of an elephant's skeleton could be used as a beginning model.

But no living species were known to resemble the complex giants and midgets that Hayden, Cope, Marsh, and their associates reclaimed in jagged remnants from the strata of the trans-Missouri West. Even the tiny horse ancestor—the British named it *Hyracotherium*, meaning "like a rabbit," but Marsh, Leidy, and Cope agreed on the name *Eohippus*, meaning "the Eocene horse" or "dawn horse"—was scientifically suspect for decades because the rabbit-size creature had four toes on each front foot and three toes on each rear foot. Also, its tooth, jaw, and leg structure indicated that it was not physically capable of grazing abrasive grasses or munching whole grains like the modern horse, but subsisted on a diet of twigs, palm buds, protein-rich bugs and mice.

As for dinosaurs, *Coryphodons*, the nightmarish "flying dragon" pterodactyls, mosasaurs, and the hundreds of other extinct species so painstakingly reborn from the boxes of rocks freighted into New Haven, Philadelphia, and Washington, the *only* living models had changed so radically through millions of years of evolution that relationships were not realized until diagnosis and resurrective reconstruction were far advanced.

Today's casual acceptance of dioramas, paintings, statues, automated models, fully "fleshed" skeletons, and detailed environmental descriptions of prehistoric life forms is due largely to those resurrectionists who, between 1870 and 1930, diagnosed, visualized, integrated, sketched, and modeled in museum and university laboratories. The achievements of these fossil surgeons, stonecutters, chemists, master smiths, and artists are more recent than the electric dynamo, the transcontinental railroad, the skyscraper, the elevator, Levi's, and chewing gum.

The pioneer work in the osteological aspects of paleontology required surgeons exceptionally skillful in identifying skeletal bits and determining their unique use by each extinct species and family during each geologic period. Caspar Wistar, Gideon Mantell, and Joseph Leidy were representative of the resurrectionists who attempted identification of prehistoric skeletons before 1850.

Except for the piecing together of the Wistar/Peale great mastodon, it appears that none of these efforts got beyond a drawing. No effort was made to reassemble Gideon Mantell's *Iguanodon*. The drawing made by Mrs. Mantell was modeled after the skeleton of a large lizard. She placed a spikelike piece of the creature's forelegs on its nose, and put what was actually part of the backbone beneath the legs.

The plaster-and-metal creatures that Owen and Hawkins created for the Crystal Palace in 1854–56 were a daring feat. Their resemblance to gigantic lizards and huge hounds with spiked elephantine hides was a good guess on the part of Hawkins and Owen. In 1869, Edward Cope erred in assembling the skeleton of a giant plesiosaur, "the ribbon reptile," at the Academy of Natural Sciences. The skull had not been discovered; Cope mistakenly installed the vertebral column "wrong end to." Joseph Leidy spotted the error and silently corrected it. But Othniel Marsh had noticed it too, and repeatedly informed audiences about it as the Cope-Marsh feud climaxed.

Between 1833 and 1840, the four volumes of Richard Owen's *Catalogue of the Physiological Series of Comparative Anatomy* were published. These, the works of Baron Cuvier, and bound volumes of "Silliman's Journal" remained the standard references for fossil detectives until Joseph Leidy's *Extinct Mammalian Fauna of Dakota and Nebraska* was published in 1869.

In 1874, Clarence King revolutionized fossil analysis when he invited Ferdinand Zirkel over from Germany to examine the rocks and fossil samples brought east by the 40th Parallel Survey and to "enrich our report with a short memoir." Zirkel was Europe's leading authority on using the microscope in the analysis and reconstruction of fossils. His "short memoir" took up 275 pages in Volume Six of the survey report, published in 1876. Titled "Microscopical Petrography," it discussed links between European and American strata and fossils, and it was illustrated with color plates that showed the similarities.

But the adaptation to the microscope was slow. Cope, Marsh, and Leidy continued to use steel tools, ranging from the stonemason's chisel to delicate probes and hand drills like those used by dentists. Since chemical seepage caused most of the skeletal parts to take on the same color as the rocks in which they were embedded, infinite patience, awareness, and sensitivity were required to extricate each bone, tooth, and segment free from the rock so that it could be integrated into the skeleton.

This profession, given the Greek-rooted name of *osteology* ("bone study") was so exacting and time-consuming that the discovery of hundreds of new species in the West necessitated dependence on diagnostic skills and drawings rather than on meticulous reassembly of entire skeletons. The genius of Edward Cope, consequently, becomes obvious. Over a period of thirty-eight years, from the time of his first publication in the Philadelphia Academy's *Proceedings* until his death in 1897, Cope published a total of 1,395 reports about his research into the Earth's dawn and "adolescence" eras—an average of 36.6 per year. Of the 3,200 vertebrate fossils that had been described in the scientific literature by 1897, Cope is credited with the discovery or first identification of 1,115. Such an outpouring simply could not wait for the intricate handwork necessary to resurrect a skeleton in the years before electric power, X rays, carbon dating, Airbrasive machines, reinforcing plastics, and other modern tools of osteology. Cope's ability to visualize skeletal structure by studying the details of a few parts enabled him to write descriptions that have sustained reexaminations by generations of scientists, including those who produced the physical resurrections of the "fleshed" creature.

Significant clues to Cope's technique were given by his devotees and professional heirs, Henry Fairfield Osborn and Charles Craig Mook, in a paper prepared for the *Proceedings* of the American Philosophical Society in 1919 (vol. 58, no. 6). Cumbersomely titled "Characters and Restoration of the Sauropod Genus Camarasaurus Cope," it describes the excavation and identification of this vegetarian dinosaur, fifty-five feet long, on the basis of "some large fossil bones" shipped from Canyon City, Colorado, to Cope in the spring of 1877 by O. W. Lucas.

After intense examination of the initial shipment from Lucas, Cope made "his original description of *Camarasaurus* and founded the genus; this description was published August 23, 1877." The creature's scientific name, Osborn and Mook explained, referred to "the cavernous nature of the centra of the cervical and dorsal vertebrae, in connection with the lateral cavities now known as pleurocoelia"; hence it became "chambered saurian," which, when Latinized, becomes *Camarasaurus*.

In October and November of 1877, the skeletal parts that had been cleaned by Cope's only assistant, Jacob Geismar, were made available to Dr. John A. Ryder, "a very able comparative anatomist" on the University of Pennsylvania faculty. From these parts, Dr. Ryder sketched a "roughly drawn, life size reconstruction" of the huge beast. This was taken to the American Philosophical Society and exhibited at the December 21 meeting.

"The reconstruction was obviously made," Osborn and Mook

decided, "after one series of bones was exposed, but before
Professor Cope had had time to give them much study. It would
not appear that Professor Cope himself seriously studied the
reconstruction, from the false arrangement of the teeth on the
malar jugal arch, and from the placing of consolidated spines
like those of the sacrum opposite the massive scapula. Twelve
to thirteen vertebrae are consigned to the neck; close to the true
number. Eighteen vertebrae are consigned to the back; eight too
many. Fifty-seven vertebrae are consigned to the tail; not far
from the correct number. A complete set of claws is consigned
to both the fore and the hind feet."

A comparison of Dr. Ryder's 1877 drawing and the fully
fleshed *Camarasaurus* Cope model that now stands in the dinosaur hall of the American Museum of Natural History provides
a succinct case history of the achievements of the osteologists
between 1877 and 1919 as well as a testimonial to Cope's
genius.

Cope's dependence on written descriptions, and his almost
disdainful attitude toward paintings, photographs, and detailed
drawings like the Ryder "reconstruction," was soon detected
and exploited by Othniel Marsh.

Cautious thoroughness had won Marsh the admiration and
legacy of his uncle George. Added to this was his reputation as
the master detective of the Cardiff Giant, the discoverer of the
"tiny horse" bones in the dirt pile of a Nebraska well dig, and
the exposer of the "kickback" system of pilfering federal
supplies intended for the Sioux reservations. The same dili-
gence was responsible for the more elaborate work in osteology
done at the Peabody Museum than Cope chose to undertake at
Haddonfield; Marsh insisted on detailed illustrations in published materials. Thus, as Osborn and Mook admitted in their
1919 paper, "Cope's references were full but accompanied by
few figures; Marsh's came later and were adequately illustrated. Marsh also issued in the publications of the United
States Geological Survey two more or less complete summaries
of characters of [dinosaurs], which were fully illustrated and
widely distributed. Consequently, they became well established

in the literature, while Cope's are still unrecognized and imperfectly known."

Details of Marsh's autocratic methods in shaping osteological techniques at Peabody Museum were recorded—with bias—in the letters and autobiographical writings of Samuel Williston. An infection in his left hand forced Williston to leave the dinosaur quarry at Como Bluff in January, 1878, and return to New Haven. He had signed a three-year contract with Marsh as a field collector at $40 a month. Marsh saw no reason why Williston shouldn't continue to earn his $1.25 a day as an osteologist at the museum while Yale physicians treated the hand infection. The laboratory routines he experienced and the ghostwriting he did that winter and throughout the decade he remained in Marsh's employ so angered him that, in 1886, he wrote to Edward Cope.

"I wait with patience the light that will surely be shed over Professor Marsh and his work," Williston wrote. "Professor Marsh did once indirectly request me to destroy Kansas fossils rather than let them fall into your hands. It is necessary for me to say that I only despised him for it. . . . The assertion of Professor Marsh that he devotes his entire time to the preparation of his reports is so supremely absurd, or rather so supremely untrue, that it can only produce an audible smile from his most devoted admirers. I have known him intimately for ten years. During most of the time while in his employ I never knew him to do two consecutive, honest days' work in science, nor am I exaggerating when I say that he has not averaged more than one hour's work per day. He is absent from the Museum fully half the time, and when in New Haven he rarely appears at the Museum till two o'clock or later and stays but an hour or two, devoting his time chiefly to the most absurd details and old maid crotchets. The larger part of the papers published since my connection with him in 1878 have been either the work of others or the actual language of his assistants. At least I can positively assert that papers have been published on Dinosaurs which were chiefly written by me."

Since Williston became the first professor of paleontology at the University of Chicago and in 1913 was awarded an honorary Sc.D. by Yale "for services in carrying on further than ever before the knowledge of the history of the life on earth in geological times," his outburst against Marsh cannot be viewed as the grumping of an incompetent. It is fair to assume that Othniel Marsh, like Louis Agassiz and many other famous scientists, became so involved in the politics and social roles he deemed essential to his position that he considered it expedient and proper to have much of his material written by assistants and published without acknowledgment. Marsh never claimed that he wrote the first drafts or assembled all of the data for his volumes. He was, like most superior editors, prone to considerable fussiness about details, phraseology, art, and—most important—publicity.

Neither Williston nor egotistical George Baur nor any of the other osteology pioneers at the Peabody laboratories ever denied that Marsh was a cantankerously precise director who "ran a tight ship," set down a vast number of rules and rigid routines, and was as clever at publicizing the results as an advance man for the Barnum & Bailey Circus.

The dig crew Marsh sent out to Canyon City, Colorado, to "bully in" on the strata where Cope's crew was discovering the pieces of *Camarasaurus* had copies of Marsh's *Quarrying Rules.* These he had compiled, discussed, re-edited, and printed in pocket-size format for his men. There were fifteen "laws," each as concise as number 5, which dictated: "Never remove all the rock from a skull, foot or other delicate specimen. The more valuable the fossil, the more rock should be left to protect it. Better send 100 pounds of rock, than leave a tool mark on a good specimen."

Cope and Sternberg's rice-paste-and-burlap-strip packaging technique for protecting bones during shipment was improved by dipping cloth strips in plaster of Paris. Later it was realized that plaster of Paris poured over the surface of a skeletal part while it was still in the rock would protect the delicate projections from injury during the rest of the extraction process.

The crew at Como Bluff averaged a ton of bones a week for

more than a decade; shipments were made to New Haven by the freight-car load. Smaller shipments arrived from Texas, Kansas, and Colorado digs. Still hungering, Marsh contracted with university instructors and graduate students to exploit the Dakota Badlands and the Judith exposures. All of these fossils had to be cleaned, sorted, catalogued, and meticulously freed from their rock coatings before the tediousness of skeletal assembly could be attempted. By 1879, it was apparent that Marsh, consciously or not, was specializing in three areas: the dinosaurs; the evolution of birds from the feathered flying reptile, *Archaeopteryx*, of the Jurassic period; and the development of the horse through the eons after *Eohippus* hid from saber-toothed cats in the Wyoming underbrush.

The Peabody laboratory staff that struggled with this deluge during the 1880's totaled eighteen and was periodically assisted—and annoyed—by Professor Dana's graduate students or an occasional congressman's nephew. One of the laboratory specialists made papier-mâché models of bones. When Marsh decided to mount his prized *Dinoceras* skeleton, all of the bones were copied in papier-mâché and strung on wire to create the display. Then the actual bones were returned to storage cabinets, since, Marsh believed, they were far too valuable to be left unprotected before curious adults, let alone small boys armed with jackknives and sticky lollipops.

Despite lost paychecks, miserly salaries, and their boss's officiousness, Marsh's staff did share his moments of glory. In 1876, Thomas Huxley visited America and accepted Marsh's invitation to inspect the collections at the Peabody. Huxley too had become fascinated by the time span of Equidae fossils being exhumed throughout Europe and was so convinced that the animal provided a splendid illustration of evolution at work that he planned to lecture about it in New York City. What began as a casual examination of the Equidae bones and skeletal drawings at the Peabody expanded into a week-long visit. When he finally got back to New York, Huxley wrote Clarence King that "there is no collection of fossil vertebrata in existence which can be compared to it. . . . It is of the highest importance to the progress of biological science that the publication of this evidence . . . should take place without delay." The New York lecture Huxley gave was rewritten to include a section praising the Peabody collection as demonstrating "the evolution of the horse beyond question and for the first time [indicating] the direct line of descent of an existing animal."

Another memorable moment occurred in 1881, when Marsh shared with the entire staff a letter just received from Down, Kent. For five years the osteologists had cleaned, glued, assembled, and wired the fragments of strange, toothed birds that Marsh and his crew had first discovered in the Kansas chalk beds in 1872–73. Thereafter, staff artists and specialists brought in from New York and Washington labored over scores of lithographs and woodcuts, and the staff concentrated on the manuscript of a volume to be titled *Odontornithes*. This volume, to be hailed as "Professor Marsh's great work on birds with teeth," was published as part of Clarence King's 1880 report on the 40th Parallel Survey. Marsh autographed copies of the report and sent them to Oliver Wendell Holmes, the famous cartoonist Thomas Nast, sundry "key" congressmen and senators, Thomas Huxley, and Charles Darwin. The acknowledgment Charles Darwin, then seventy-two, and within a year of his death, sent from Down, Kent, read in part: "I received some time ago your very kind note . . . and yesterday the magnificent volume. I have looked with renewed admiration at the plates, and will soon read the text. Your work on these old birds and on the many fossil animals of North America has afforded the best support to the theory of evolution which has appeared within the last twenty years."

By 1881 such praise from the father of the theory of evolution was more welcome to the osteologists than a raise in pay. Marsh meanwhile had become involved in a political wrangle that was arousing more anger in his staff than either the delinquent pay checks or crotchety editors. It began with the death of Joseph Henry in 1878.

In 1878, Henry was serving as president of the National

Academy of Science. That same year, Marsh was elected vice-president. There were some who suggested that the publicity Marsh received during the tussle with Christopher Delano and the Bureau of Indian Affairs influenced his election. Upon Henry's demise, Marsh became acting president. A month later, a congressional committee investigating government expenditures asked the Academy an inane question: what reorganization is necessary to end the duplication of labor and competition among the four geological surveys?

Perhaps the question was introduced by a congressman intent on demonstrating to his constituency his vigilant guardianship of the public purse. It had been obvious to everyone else that the instructions given to Hayden, King, Powell, and Wheeler were so vaguely grandiose that some overlap was inevitable. All of them, in fact, benchmarked, dug, exhumed, and elaborately reported on huge patches of each other's territories. Powell explored *down* the Colorado River; Wheeler was defeated trying to explore *up* it. Hayden surveyed and mapped from Colorado Springs to Santa Fe; so did Wheeler. Powell explored and surveyed the Green River; Hayden explored and surveyed most of the Green River. King's orders were to explore "as far east as the 105th Meridian," which bisects the mountains between Laramie and Cheyenne, Wyoming; Hayden's orders were to survey the Wyoming Territory, including large chunks of the areas assigned to both King and Powell. Obviously, the congressional committee's question should have been: how can we be more precise in our instructions?

Marsh appointed a committee to investigate the situation and make recommendations. The committee, with questionable alacrity, voted to urge consolidation of all four surveys into the United States Geological Survey, which would be administered by the Department of the Interior. Edward Cope, also a member of the National Academy of Science, cast the only dissenting vote. The congressmen, eager to cover the tracks of their own foolishness, hurriedly turned the recommendation into a law.

A struggle ensued for the post of director of the U.S. Geolog-

ical Survey. Scientists throughout the country agreed that the
appointee "must not be Wheeler or any other creature of the
Army." Hayden was confident that he would be chosen, but he
did send letters to university officials and legislators asking
them to support his appointment. Marsh campaigned vigorously for King, outwardly basing his position on the fact that
"we were classmates at Sheffield Scientific," but with the hidden
motive of knocking out Hayden and thereby cutting off Cope's
best source of publication.

Powell lobbied for Hayden for a few weeks, then suddenly
opted for the Marsh-King combination and urged his supporters
to switch too. Clarence King was appointed the first director of
the U.S. Geological Survey in 1879. Hayden stayed on the
federal payroll as director of the Montana Division of the
survey, but without field assignments. Cope was out.

By a not-so-strange coincidence, agreement had been reached
that the final report of King's survey (volume 7, 1880) would
include, as a supplement, *Odontornithes, a Monograph on the
Extinct Toothed Birds of North America*, by Professor Othniel
Marsh.

Cope eagerly explored the West during most of this bureaucratic poker game. He inspected the Marsh quarry at Como
Bluff and made such a favorable impression that the British
teacher/painter Arthur Lakes wrote in his journal, "The mon-
strum horrendum Cope has been and gone, and I must say that
what I saw of him I liked very much. His manner is so affable
and his conversation very agreeable, I only wish I could feel
sure he had a sound reputation for honesty." (It seems probable that Marsh was Lakes's *only* source of information about
Cope's "reputation for honesty.")

Moving farther west, Cope began a fossil hunt in the Silver
Lake region of Oregon. Regretfully turning east in late September, he decided against a search through the Wasatch Mountains' glimmering barrier of the Great Salt Lake valley, and
instead briefly checked on the dig he had ordered Lucas to
begin "out of revolver range" of the Marsh quarry at Como. He

also visited the dig at Canyon City, then hired a cook and wagon and jolted back into the Rockies' Front Range to search for Paleocene and Eocene fauna. He did not learn all of the details of the Marsh coup until he reached Philadelphia in mid-November.

But things were not too bad, he decided after long discussions with his wife. The year before he had bought an interest in a moribund magazine called *The American Naturalist*. The investment was made because James Dana, now editor of "Silliman's Journal," was also professor of geology at Yale and Marsh's department chairman. Naturally, Marsh and the Peabody Museum could exert more pressure on Dana for use of news articles and "discoveries" than Cope could. If *The American Naturalist* could be perked up a bit, Cope reasoned, it could provide as much publication space as the Hayden survey reports—and far more quickly. Furthermore, the government was still obliged to publish Hayden's 1879 report, which was scheduled to include Cope's cherished manuscript on "Vertebrata of the Tertiary Formations of the West." He had worked on it—spasmodically, to be sure, but nevertheless worked on it—since 1872. It promised to fill thousands of pages. Indeed, matters did not look at all bad to Cope.

Bureaucratic backbiting is the plague of any government, and Washington, D.C., had its share of people who were all too ready to engage in petty malice. The records allege that Clarence King resigned as director of the U.S. Geological Survey in 1881 "to enter private employment," but the relationship between government record and truth is frequently no more than a polite nod in passing. John Wesley Powell became the director; Othniel Marsh was promptly appointed U.S. Geological Survey paleontologist.

Marsh's salary as paleontologist was four thousand dollars a year. He deposited all of it in the Peabody Museum's account. But, as home of the Paleontologic Division of the USGS, the Peabody was granted other employees on the federal payroll. Marsh and Powell agreed that the USGS would provide funds directly to Professor Marsh, to be disbursed at "his discretion." The payments averaged fifteen thousand dollars a year during the ten years of Marsh's term as USGS paleontologist.

The sum seems minute in relation to the inflated economy of the 1970's. But in July, 1882, Samuel Williston, recently married and soon to become a father, signed his third contract with Marsh. In it he agreed to work for Marsh "for five years, forty hours a week and one month's vacation each year, for $1,500 annually."

The salary of John Bell Hatcher, the Peabody's most skilled collector, was just as meager. Between 1888 and 1891, Hatcher discovered enough strange bones in Montana's Judith River country and in Wyoming to validate the existence during the last few million years of the Cretaceous period of a nightmarish three-horned dinosaur. This vegetarian, with an average length of eighteen feet, had evolved not only horns but also a shield of bone atop its skull. The shield and skull equipped it to compete with *Tyrannosaurus*, the ugliest and most vicious of all carnivorous dinosaurs. Marsh gave the skeleton the name of *Triceratops*—and boosted Hatcher's salary to two thousand dollars a year! Such a salary scale suggests that the U.S. Geological Survey underwrote a most generous percentage of the osteological research and dig-crew budgets of the Peabody for at least a decade.

But the astute and tough John Wesley Powell obtained compound returns on this investment of public funds. The bones discovered and retrieved by Marsh's dig crews and laboriously cleaned and analyzed by the Peabody osteologists enabled Marsh to be credited with adding 536 genera and species of fossil vertebrates to the dawn history of North America. This was less than half as many as Cope identified, but it exceeded the 375 credited to Leidy, and it did include three of the most popular museum attractions: nineteen genera of dinosaurs, the evolutionary cycle of the horse, and the still mysterious pterodactyls and the toothed birds. In addition, seven freight-car loads of the specimens extracted from the Kansas chalk beds, the Judith region, Canyon City, and Como Bluff were given to the Smithsonian's U.S. National Museum, and Marsh volunteered his services to the Smithsonian as a lecturer on osteological procedures, display techniques, organization of storage bins, and cataloguing.

Meanwhile, the osteologists at the Peabody were busy upsetting old folk tales and creating new ones. The most cherished myth they "clobbered" concerned those "Noah's raven" footprints in the Connecticut valley sandstone that had proven so intriguing in the first half of the nineteenth century to New England divines and to President Edward Hitchcock of Amherst. Quarrymen uncovered, and Marsh bought for the Peabody, the bones of three dinosaurs entombed in Connecticut valley Triassic rock. One of the skeletons, called *Anchisaurus*, was cleaned, assembled, and cautiously arched on wire until it was tilted into a leaf-nibbling position. It became one of the Peabody's permanent exhibits of Connecticut history.

In 1882, the year that the iguanodont skeletons discovered in the Fosse Sainte-Barbe coal mine first went on display at the Royal Museum in Brussels, a young French diagnostician named Louis Dollo joined the Royal Museum staff and began publishing papers about the structure and habitat of the *Iguanodon* as revealed by the Bernissart deposits. The Dollo papers, along with Cope's brilliant deductions and Marsh's confidence in the skills of his own staff, led Marsh to the conclusion that the Peabody could assemble and display America's first skeleton of a *Brontosaurus*, one of the mammoth vegetarians of the late Jurassic. The task was begun in about 1890 and today the skeleton, which stands in the great hall of the Peabody Museum, is an impressive memorial to Marsh.

While his political sophistication and the competence of his dig crews and osteologists made Marsh world famous, wrong guessing pushed Edward Cope closer and closer to a desperate act. When Marsh was appointed paleontologist of the U.S. Geological Survey, Cope decided that his only alternative would be to outcollect and outwrite Marsh and his entire corps of specialists. To do this he would need far more money than his inheritance could provide. He began to invest in Western mines, interspersing his fossil hunts to the Rockies, Texas, and Oregon with visits to gold and silver "strikes." But the claims petered out and the mines closed. By 1888 Cope was impoverished and barely able to pay his only assistant, faithful Jacob Geismar.

Cope had long before exchanged his Haddonfield property for two four-story houses at Pine and Twenty-first streets, near midtown Philadelphia. This move, he reasoned, would provide his daughter, Julia, with more social facilities and urban conveniences during her years at Bryn Mawr College and would also put him within pleasant strolling distance of the Academy of Science, the American Philosophical Society, the Union League Club, and other beloved haunts. Cope's family lived in one of the buildings; the other served as his laboratory-library-mausoleum. Many of the boxes of fossils from the West were still stored, unopened, in basement rooms. He continued to turn out dozens of articles and, by skimping on travel and sometimes on household expenses, managed to keep *The American Naturalist* going.

By 1888 he was almost as bitter toward John Wesley Powell as he was toward Marsh. In 1883 he had finally completed the manuscript and drawings for Volume One of *Vertebrata of the Tertiary Formations of the West*. Powell accepted it as a "supplement" to Hayden's 1879 Report, but scolded Cope about its 1,000 pages and its 1340 plates. "The publication funds for Hayden's reports are now exhausted," he announced; thus it would be impossible to publish the second volume Cope had planned.

The reviews of Volume One of *Vertebrata* were glowing. The book became so essential to lecturers, graduate students, and osteologists that it was given the nickname "Cope's Bible." But Powell, knowing that Volume Two was still only in outline, remained adamant about not publishing it. During 1885–87, Cope made frequent trips to Washington where he attempted to persuade congressmen to "order" Powell to consent to the publication of Volume Two. He did not have Marsh's suavity, succeeded only in gaining a reputation for being a nuisance, and irked Powell into greater dependence on Marsh.

The financial horizon cleared slightly in the fall of 1889, when University of Pennsylvania trustees responded to the urgings of Professor Emeritus Joseph Leidy and offered Cope a professorship in geology and mineralogy. Cope's lectures were lively, and he attracted many students. Characteristically, Cope decided that the available textbooks were inadequate. He cleared away desk space at the Pine Street study, secluded himself there for a few weekends, then bustled his new "and quite thorough" textbook off to the printer. He seemed happier than he had been since the Judith adventure with Sternberg.

One December morning a letter from the Department of the Interior was dropped into the Pine Street mailbox. It was signed by Secretary of the Interior John Noble and stated that Cope was required to turn over to the Smithsonian all of the Cretaceous and Tertiary fossils he had collected during his years as paleontologist of the Hayden survey.

Marsh, as Cope knew, had recently caused such a rule to be inserted in USGS contract forms. But this was a demand for all of the fossils Hayden had shipped to Philadelphia between 1868 and 1879. Most of the osteology on them had been done gratis by Cope, with the understanding that they would become his property. If he submitted to Noble's demand, Cope reasoned, most of the collection would wind up at the Peabody. The time had come to expose the whole sordid situation.

William Hosea Ballou had worked for a while as an editor of *The American Naturalist* before moving to New York City. There the efforts of Joseph Pulitzer to stimulate newsstand sales of his *New York World* were drawing more and more publishers into what eventually became known as "yellow journalism." Ballou finally entered the fray as a reporter on Bennett's *New York Herald*.

Cope called Ballou to Philadelphia, poured out all of his bitterness about Marsh and Powell, showed him the grumping letters he had received from Williston and other Marsh associates, and supplied the names of people who might wish to air their feelings.

Hayden had died in 1887. Leidy sadly shook his head and whispered, "No comment." But when Ballou posed as a scientist "investigating the Geological Survey's expenditures," he obtained enough reluctant allegations to convince the *Herald*'s editors that the "exposé" rated feature treatment in a Sunday issue. The boil burst on January 12, 1890, under the headline:

SCIENTISTS WAGE BITTER WARFARE

Prof. Cope of the University of Pennsylvania brings serious charges against Director Powell and Prof. Marsh. Corroboration in plenty.

The muckrakers worked through the mess for a month. George Baur hurriedly resigned from the Peabody staff, fled west, and joined the faculty of the new University of Chicago. Several of the professors Ballou had interviewed claimed that they had not talked "for publication" and had been "grossly misquoted," anyway. Powell wrote a dignified defense of USGS policy, praised Cope for his "valuable work for science," and concluded: "If [Cope's] infirmities of character could be corrected by advancing age, if he could be made to realize that the enemy which he sees forever haunting him as a ghost is himself . . . he could yet do great work for science."

Marsh opened his defense by taking the train to Philadelphia to console the University of Pennsylvania's president and trustees in this "shame that has befallen you." After insinuating that "poor Cope" might be cracking up, he volunteered his services in locating a "more substantial scientist" to replace Cope on the faculty. Then, in the January 19 issue of the *Herald,* Marsh filled an entire page with denials of Cope's charges and a wolfish recital of Cope's shortcomings and diagnostic errors.

When it was all over, the situation had not changed very much. Cope retained both his professorship and the Hayden fossils; Marsh and Powell retained their appointments. Ballou returned to the drudgery of reporting on WCTU conventions and Suffragette parades.

The fundamentalists finally bagged Marsh. Congressman Hilary Herbert of Alabama was the instrument of their wrath. When the 1893 budget of the U.S. Geological Survey came up for discussion before a House committee in 1892, Congressman Herbert advised his home-town papers to stand by. His well-rehearsed punch line was "Birds with teeth!" followed by a grimace, the waving of a book, and finally an anguished shout: "That's where your hard-earned money goes, folks—on some professor's silly birds with teeth." The book Herbert was waving was the privately printed edition of Othniel Marsh's *Odontornithes*, and the message was clear: such high-falutin nonsense should not be underwritten by the tax receipts of the American people.

"Birds with teeth" was a catchy phrase. It gained such popularity in the House that "Big Bill" Stewart of Arizona and a few other senators joined the witch hunt. Marsh was a Darwinist; Louis Agassiz's son, Alexander, was still pontificating for the fundamentalists at Harvard, and so joined the chorus with well-chosen, mellifluous quotes against "the Godless." Powell was finally forced to telegraph Marsh: "Appropriation cut off. Please send your resignation at once."

That was it. Cope taught happily at the University of Pennsylvania for five more years, went on digs to the Dakota Badlands in 1892 and 1893, and, on April 12, 1897, died on a cot in his Pine Street study. His last request was to be buried beside Joseph Leidy, who had died in 1891.

Marsh had spent all of his uncle George's legacy on the Western digs and the Peabody osteologists, so was forced to accept a modest salary from Yale. He lived alone in an eighteen-room brownstone mansion near the campus, puttering about each day in his orchid hothouse and flower beds. In 1896 he published his greatest work, *The Dinosaurs of North America*. Early in March, 1899, Marsh contracted pneumonia. He died on March 18; his bank balance totaled less than a hundred dollars. All of the fossils at the Peabody, as his will stipulated, became the property of Yale University.

The nineteenth century had opened with the discovery and

assembly of the Peale mastodon skeletons and closed with most of the world's literate population prepared to accept the substantial fossil evidence that the Earth was hundreds of millions of years old, and that evolution was a continuing process. Much of this evidence came from the discoveries and osteological advances made by Edward Drinker Cope and Othniel Marsh. Driven on by the egotism of their rivalry, they made huge contributions to American civilization—some of them subtle, most of them obvious in their social and economic applications. In addition to being discoverers of the oldest West, these men were heroes of the new, Wild West. An era of glory ended with their passing.

18 | "Declare, if Thou Hast Understanding"

He carried many profound discoveries of a scientific epoch into homes and schools throughout the world, changing "dinosaur" from a highbrow to a household word and making Mesozoic dragons almost as familiar to children as the creatures of Noah's Ark.—Robert C. Murphy, *Reminiscences*, 1947

BY 1897 the torrents of anthracite and petroleum compounds spewing from chimneys and smokestacks were tinging morning mists off the Delaware and Schuylkill rivers a funereal grey. Hence the bars of sunlight slanting through the damask draperies of 2102 Pine Street in Philadelphia on the afternoon of April 14 were the same ochre sheen as the surface clays of Bridger Basin and Como Bluff. The sunlight bounced off the legs and drawer pulls of Edward Drinker Cope's desk, then reflected upward to silhouette a casket atop the desk and rows of men and women seated on ladderback chairs. The only sounds for half an hour had been the muted rumble of trolley cars, the clop of horse drays, the occasional basso of a steamboat whistle, the falsetto response of tugboats.

A chair creaked. A stocky figure rose, rustled the pages of his pocket Bible open to the thirty-eighth chapter of the Book of Job, and began reading in the clipped accent of a Yankee aristocrat: "Where wast thou when I laid the foundations of the earth? declare, if thou hast understanding. Who hath laid the measures thereof, if thou knowest? or who hath stretched the line upon it? Whereupon are the foundations thereof fastened? or who laid the cornerstone thereof; When the morning stars sang together, and all the sons of God shouted for joy?"

Dr. Henry Fairfield Osborn of the American Museum of Natural History gently closed his Bible and sat down. A woman sobbed. Hands brushed away tears. The silence of the Quaker funeral service resumed. A half hour later, Dr. Osborn became one of the pallbearers for Edward Drinker Cope's final journey to a resting place beside his mentor and friend, Joseph Leidy.

Dr. Osborn's participation in Cope's funeral was one of the most significant acts of a distinguished lifetime. He became Cope's heir in paleontological achievement, as well as his biographer. Only three weeks before, Osborn had introduced Cope to the professional illustrator Charles Knight. He had persuaded Cope to give Knight "accurate details" for paintings of dinosaurs and other prehistoric creatures. Cope, with his extensive knowledge of the bone structure and musculature of extinct mammals, was able to provide Knight with essential data for visualizing the living appearance and environmental background of dinosaurs, *Eohippus,* saber-toothed tigers, giant sloths, mastodons, mosasaurs, and other "dragon" beasts of prehistory. Thus Cope, Knight, and Osborn pioneered those depictions of the dawn eras' life forms that have become standard in museums, advertisements, movies, television, books, statues, postage stamps, and encyclopedias.

The friendship between Cope and Osborn began in 1877. Like Cope, Osborn was a child of wealth. His father reorganized the Illinois Central Railroad and amassed a fortune through its land grants and monopolistic freight rates. An aunt married Junius Pierpont Morgan, Sr.; although the couple divorced, Morgan remained "Uncle Pierpont" to Osborn and became his most dependable "angel" in financing expeditions and publications. One of the Osborns' favorite properties was the replica of a Rhineland castle built on a bluff of the Hudson highlands, opposite West Point and ten minutes by boat from the estate where the former Mrs. Morgan grew orchids and other exotic plants in a huge, lavishly staffed greenhouse.

Osborn and William Berryman Scott decided to celebrate their graduation from Princeton in June, 1877, by going to Bridger Basin and digging for Eocene fossils. Their geology professor urged them to see Cope, who lived only an hour away from the campus, before they went.

Cope was absorbed in analyzing the first shipments from the Canyon City dig, so he gave Scott and Osborn only monosyllabic replies to their questions. Undaunted, they went to Fort Bridger anyway, then hired a wagon and drove out to the dig being financed by Othniel Marsh. They were brusquely ordered to "move on, because Professor Marsh permits no amateurs at his digs." Stubbornly, the twenty-year-olds combed the hillsides until they recognized an Eocene outcropping; there they found excellent skeletons. That fall they returned to 2102 Pine Street with several of the bones and asked Cope to identify them. When Cope learned that the Marsh crew had exceeded him in rudeness, he became loquacious, identified the bones as those of a titanothere, and volunteered to sketch this elephantine giant.

Subsequent sessions with Cope persuaded both Osborn and Scott to specialize in paleontology. Teased on by a fascination with titanotheres that would, half a century later, sire a two-volume, sixteen-pound work titled *The Titanotheres of Ancient Wyoming, Dakota and Nebraska,* Osborn studied histology and anatomy at Bellevue Medical College in New York City. In 1879, again with Scott, he sailed to England to learn the intricacies of comparative anatomy and embryology in the small, elite groups tutored by Francis Balfour and Thomas Huxley. By 1883, he was back at Princeton, as professor of comparative anatomy; Scott was already there as an associate in paleontology.

The friendship with, and tutoring by, Cope continued. Othniel Marsh contributed to the relationship by assuming that Osborn and Scott were paid spies for Cope. He ordered the osteologists at the Peabody to be extremely cautious about giving information to "those snoops from Princeton." After a particularly frustrating exchange of letters, Osborn and Scott decided to go to New Haven and ask permission to examine Eocene and Permian specimens. The Peabody's doorman took their calling cards and disappeared. Fifteen minutes later a young man in a smock entered the anteroom, introduced himself as their guide, and expressed Professor Marsh's regrets at being unable to conduct "his distinguished colleagues from

Princeton" himself. The tour was brief; the guide's answers to
questions were evasive. But Osborn and Scott were too intrigued by "the shadow" to concentrate on the few bones being
taken out of cabinets for them. Some twenty-five feet behind
them, tiptoeing in carpet slippers, was the rotund, bearded
figure of Othniel Marsh. Hiding behind pillars and packing
cases, he peered at each bone being exhibited and gave head-
shakes and hand signals to the guide. With head bent and
hands clasped behind his back, he waddled wordlessly past
Osborn and Scott while they were exchanging final pleasantries
with their guide.

Urged on by "Uncle Pierpont" and other wealthy friends of
the Osborn family, the trustees of Columbia University and the
American Museum of Natural History approached Osborn in
1891 with a joint offer. Almost a century after Dr. Samuel
Latham Mitchill initiated its natural science lectures and museum, Columbia University had decided to found a department
of biology; the trustees offered Osborn the challenge of organizing it. This would be part of an integrated exchange program
in which Osborn would also set up a department of mammalian
(later vertebrate) paleontology at the American Museum of
Natural History, now administered by the eminent paleoichthy-
ologist Bobb Schaeffer. All of Columbia's graduate studies in
paleontology would be transferred from the university's Morn-
ingside Heights campus down to the museum's rococo gloom
at Seventy-ninth Street and Central Park West, as would
Columbia's large collection of fossil fish, which had somehow
become trapped in its School of Mines.

Osborn accepted the challenge and, with a brusqueness that
rivaled Cope's, began the physical and staff changes that would
establish the American Museum of Natural History as the giant
of twentieth-century museums. Although the museum had purchased James Hall's collection of New York, Michigan, and
Dakota fossils in 1875, its curators remained indifferent to the
fossil giants recovered from trans-Missouri strata by Hayden,
King, Cope, and Marsh. But Osborn, like Charles Willson
Peale, possessed a showman's instinct. He decided that giant

fossils would lure both visitors and sponsors. Moreover, thanks to the osteological innovations of Cope and the Peabody's technicians, he knew that it would now be possible to put scientifically accurate "skins" on the skeletons and exhibit them in dioramas faithful to the geological eras in which they had lived.

Cope was Osborn's confidant and advisor during these planning stages. He recommended one of his former assistants, Dr. Jacob Wortman, as a "remarkable" osteologist and collector. During 1893 and 1894, cautious negotiations began for the museum's purchase of Cope's fossil collection, "if and when, of course, you ever decide to part with it."

Three other young men who would become famous bone hunters were put on the museum's payroll soon after Osborn moved in. Samuel Williston, who had returned to the University of Kansas as its professor of geology, recommended his star pupil, Barnum Brown. Osborn watched the graduate work of William Diller Matthew at Columbia for several months, then recruited him. Walter Granger had hiked down from his native Vermont as a seventeen-year-old in 1890 and applied for "a job where I can learn taxidermy." He was hired as janitor for the taxidermy shop and later was given the not altogether pleasant job of skinning and preparing the pelts of animals who died at Central Park Zoo. He did so well that, in 1894, Osborn sent him west to assist Dr. Wortman on a dig.

In 1895, Cope finally consented to sell his cherished fossils to the American Museum of Natural History, and he agreed to Osborn's bargain-basement offer of $32,000. Osborn saw no reason to argue the sale with the trustees; Pierpont Morgan had set up a drawing fund for paleontological purchases by Osborn that would never be revealed to the museum's auditors. William Diller Matthew was sent down to Pine Street to supervise the packing and removal. He found dozens of crates and bone bundles that had never been opened. Cataloguing and analyzing the Cope collection occupied most of Matthew's office hours for the next decade.

Osborn believed, and most sociologists agreed, that New York would become the world's largest city. Therefore, he concluded, New York should have the world's largest collection of dawn-era fossils. Moreover, nobody knew when the supply would be exhausted by the eager digs of museums, universities, and rock enthusiasts. He suspected that there were still quantities of Mesozoic skeletons in those Como Bluff quarries that Marsh had been forced to abandon when the fundamentalists forced him off John Wesley Powell's U.S. Geological Survey payroll. The day after Cope's funeral, Osborn was back at his desk dictating instructions for a Como Bluff expedition. His memorandums, typed on letterhead stationery and formally signed, were delivered by a secretary to the office, one hundred feet away, occupied by Wortman, Brown, and Granger.

The 1897 expedition was, as *The Morning Telegraph* would have expressed it, "a real bust." Williston, Reed, Arthur Lakes, and the other Marsh collectors had hacked and pried the quarries clean, except for a few worthless splinters. Osborn went out to Como himself and ordered Wortman and Brown to join him in a search toward the Cooper Lake badlands, then west toward the Medicine Bows. They found two skeletons in moderately good condition and spent the rest of the summer extracting and packaging them.

Still adamant, Osborn ordered the same team back to Como in the late spring of 1898, with instructions to explore north up the Medicine Bow anticline. The search seemed futile. At dusk on June 11, Dr. Wortman sighted a cabin on a hilltop a mile or two away. He decided the cabin would be their benchmark for the next day's search, and ordered the tents pitched.

As the team fanned out across the plain next morning, each person peered and poked for evidences of fossil bones, then turned and headed up the hill toward the cabin. Wortman's shout was echoed by Granger's. Each had been suddenly confronted by giant thighbones and rib sections, black and twisted. Then Granger stared up at the cabin, shouted again, and began running. The cabin was built entirely of dinosaur bones, cemented together with dollops of adobe and sod.

A shepherd had viewed the hillside strew of "old bones" as a godsend in the treeless countryside, and had built a winter shelter from them. Bone Cabin Quarry was one of the richest paleontological treasure troves on the continent. Two years later, the veteran Bill Reed sold Osborn the digging rights to another dinosaur cemetery he had discovered near Rock Creek, fourteen miles away. Between 1898 and 1903, Bone Cabin and Reed's Quarry R yielded 493 Jurassic dinosaur specimens, weighing approximately 180,000 pounds.

Edward H. Harriman had gained control of the Union Pacific Railroad. Harriman was not nearly so gracious about fossils and the wonders of science as his predecessors had been. Consequently, the bones rode east only after "Uncle Pierpont" agreed to pay the freight bill. (This was, of course, prior to the Harriman-Morgan partnership in the Northern Securities Association, which was declared a trust and was "busted" by the Supreme Court.)

While Granger, Brown, Matthew, and Wortman sharpened their skills at Bone Cabin and Quarry R, Osborn experimented in New York with visual techniques that would take museums out of the skeleton-on-a-wire rut and bring them back to Charles Willson Peale's dream of environmental realism. The experiments had begun in the museum's taxidermy shop in 1891, when Walter Granger was still "that young janitor and pelt skinner." There Charles R. Knight, a seventeen-year-old art student from Brooklyn, began sketching the anatomies of carcasses from the Central Park Zoo and comparing them first with fossil skeletons being assembled in laboratories down the hall, and then with the live animals at the zoo. By 1894, Wortman was helping Knight with water-color portraits of *Elotherium* and other prehistoric mammals.

There was little precedent for these efforts to depict extinct species "in action." A few of the pencil sketches in Cope's notebooks showed horned lizards and "dawn emperors" in lifelike stances; over the drawings, Cope had scribbled such self-critical notes as "The horns and head are rather too large." Arthur Lakes, the Oxonian schoolmaster/clergyman of Golden City, Colorado, loved to paint landscapes. During his years as Marsh's representative at the Como Bluff dig, he irked Bill

Reed by spending more time painting than digging. But Lakes's scale drawings of the monster bones as they were cautiously chipped from the strata, his depictions of a bone quarry after a blizzard, and his imaginative sketches of pterodactyls, mosasaurs, and dinosaurs diving, grimacing, and fighting in blue-green shorescapes were of little use to the osteologists. Except for Cope's sketches, Arthur Lakes's paintings, and the pioneering work of Waterhouse Hawkins, all the illustrations of paleontology's ancient world were, like the drawings Peale made for Dr. Michaelis, merely depictions of bones and teeth.

By 1894, Knight had learned enough about the proportions and musculature of dinosaurs, giant sloths, titanotheres, saber-toothed tigers and the multitoed ancestors of the horse to be able to model lifelike figures of them in clay. "From evidence at hand," Clayton Hoagland reported in his splendid article, "They Gave Life to Bones" (*Scientific Monthly*, February 1943), "he was able to draw conclusions about the feeding habits of the animal, and its attitudes, as shown by the position of the joints and the angles of feet and limbs. Then, to get the solidity of three dimensions on paper and to reproduce shadows of the animals on the ground, the models were placed in sunlight, and the paintings made from them."

Now convinced that the museum was on the right path to scientifically legitimate popularization of prehistoric life, Osborn talked "Uncle Pierpont" into giving him another research grant and began to team osteologists with Knight in preparing skeletons in life poses. "While in New York in 1896," Hoagland reported, "Professor Schuchert of Yale saw the skeleton of a brontothere mounted by Adam Hermann in accordance with Osborn's ideas. When the method was proposed to G. Browne Goode, then director of the National Museum in Washington, D.C., however, Schuchert was informed, in somewhat contemptuous tones, that he had not seen fine paleontology in New York, but fine art. In Washington, as in many other museums, the collections of skeletons of fossil vertebrates were left on the shelves for the exclusive [awareness] of paleontologists—until the movement promoted by Osborn spread."

Shortly after the momentous March, 1897, discussions between Knight, Osborn, and Cope in a New York hotel room—Cope wrote his wife that Knight "is very original in attitudes"—Osborn told Knight to revise his clay figurines to comply with Cope's suggestions and then have molds made of them. A modest supply of bronze casts was made from the molds. The models were also photographed; postcards and stereopticon slides were prepared. All these items went on sale at the museum in 1898. An illustrated catalogue inviting mail orders was sent out with membership solicitations. "The true modern spirit in which to study a fossil vertebrate," Osborn emphasized, "is to imagine it as living, moving, walking, swimming or flying, begetting its kind. . . . We have secured complete skeletons in the place of fragmentary parts," he continued, and, since the organs of sight and smell had been studied as part of "fossil psychology," it was possible to consider these beasts as "thinking, inasmuch as they obviously recognized both enemies and prey" and devised their methods for survival.

R. Bruce Horsfall and Erwin S. Christman joined Knight in the museum's art department. Horsfall was a twenty-eight-year-old Iowan who had studied art in Munich and Paris and exhibited his works at Chicago's massive Columbian Exposition in 1892–93. Christman was a brash fifteen-year-old when he first approached Wortman, declared an exuberance for drawing "things about animals," and was encouraged to "go talk with Mr. Knight." After Knight tested the youth by asking him to make a wash drawing of titanothere teeth, introductions to the Art Students League were written and a desk assigned for "playing about with pencils and clay."

Century Magazine became the pioneering vehicle for the environmental and action paintings that would strongly influence people's images of leaping dinosaurs and mosasaurs, saber-toothed tigers challenging hulking, dagger-fingered sloths at the edge of a tar pool and the ludicrous moose-faced, walrus-tusked, elephant-bodied profiles of uintatheres as they chewed their palmetto-and-vine-twig cud. Osborn used several paintings by Knight to illustrate laudatory articles about Cope, which eventually formed the core of his definitive biography, *Cope: Master Naturalist*, proudly published in 1931 by Princeton University Press (where years before Cope had been turned down for a professional post because of "radical" views). The Osborn-Knight articles were published during 1896 and 1897. In its June, 1898, issue, *Popular Science Monthly* leaped aboard Osborn's mass-appeal band wagon with a sensational article about "The Serpentlike Sea Saurians." It was illustrated with pen drawings by Edward Cope, Othniel Marsh, and Samuel Williston and was dominated by a full-page scene by J. Carter Beard depicting mosasaurs, turtles, a plesiosaur, and several snapping "bulldog" *Portheus* engaged in a frothing free-for-all in "the great Cretaceous ocean." By 1898, the Pulitzer-Hearst circulation struggle was in full fury; some historians contend that the Spanish-American War was one of its grimmer by-products. The razzle-dazzle of prehistoric creatures intent on killing one another was almost as spectacular an art theme as Admiral Dewey's squadron blasting the Spanish defenses of Manila Bay or Colonel Theodore Roosevelt charging up San Juan Hill. Both Hearst and Pulitzer Sunday supplement editors saw the Beard drawing and summoned their staff artists.

Erwin Christman's skills matured during his studies at the Art Students League and the American Academy of Design, where he showed extraordinary sensitivity as a sculptor. He made the models that were photographed to illustrate Osborn's book *The Age of Mammals in Europe, Asia and North America;* then, under the coaching of Osborn and William K. Gregory, sculpted the life-size heads of the rhinolike titanotheres that, during April, 1912, were placed on permanent display in the Hall of Vertebrate Paleontology. Later, he and Gregory prepared a complete restoration of titanotheres and, for the first time, reproduced their intricate musculature.

Christman's most notable achievement took several years and was completed shortly before his death, in 1921. This was a complete restoration of *Camarasaurus*, one of the Jurassic giants that Cope discovered and identified in 1877–78. Christman made scale drawings of each of the giant's bones. These drawings were glued on board and cut out. Christman labored with the vast puzzle until he had achieved a realistic pose. Only then did he sculpt the model.

Horsfall worked just as purposefully in the Seventy-seventh Street studios and at the museum that William Berryman Scott, Osborn's classmate and lifelong friend, was developing at Princeton. He illustrated articles about paleontology for *Mc-Clure's Magazine*, did realistic backdrops for skeletal exhibits in the Hall of Vertebrate Paleontology, illustrated Scott's *History of Land Mammals in the Western Hemisphere*, and "resurrected" the fossils collected during 1899–1902 by the Princeton Museum's Patagonian expedition.

Osborn's campaign to impart realism to exhibits was greatly assisted by Charles Sternberg and his sons during the fall of 1908. Sternberg, like Cope, was a determined individualist. He disdained the regimentation and status struggles of government or university employment. After Cope lost his fortune in mining investments, Sternberg became an independent contractor. Natural-history museums were being set up at universities, state capitals, and in most major cities. They wanted fossils from the West; Sternberg provided excellent specimens, carefully exhumed, packaged, and delivered on schedule. Consequently, he earned a good living and gave his three sons, Charles, George, and Levi, such arduous training in the field that each became a famous collector or curator.

During the summer of 1908, the four Sternbergs worked Cretaceous beds on the Cheyenne River headwaters of eastern Wyoming, sixty miles from Lusk. The nearest stores were in Lusk. When their food supplies had dropped down to the last bag of wrinkled potatoes, the elder Sternberg decided that he and Charles Jr. would drive into Lusk for flour, salt pork, and perhaps a few vegetables. George and Levi would continue the search for a "lucky break" in what had been a most disappointing summer. The day after the wagon left, George discovered the rib ends of a dinosaur jutting from a sandstone cliff. "By the evening of the third day," he recalled years later, "we had traced the skeleton to the breast bone, for it lay on its back with the ends of the ribs sticking up. There was nothing unusual about that. But when I removed a rather large piece of sandstone rock from over the breast, I found . . . a perfect cast of the skin impression beautifully preserved. Imagine the feeling that crept over me when I realized that here for the first time a skeleton of a dinosaur had been discovered wrapped in its skin. That was a sleepless night for me!"

Five years later, on the Red Deer River, north of Calgary, Alberta, the Sternbergs climaxed a successful summer by retrieving a *Chasmosaurus* "wrapped in its skin impressions."

These discoveries contributed some new data about the skin and cuticle of the "terrible lizards" that by and large validated the analyses of Cope and the Peabody and American Museum osteologists.

A new method for validating the age of fossils and their encompasing strata was first proposed by B. B. Boltwood in an article published in *The American Journal of Science* in 1907. Dr. Boltwood based his method on the amount of lead generated in a radioactive mineral by atomic disintegration. Although many scientists sneered at the suggestion for years, repeated experiments validated the preciseness of what came to be known as uranium-lead dating. "It has grown in versatility," Professor Adolph Knopf of Yale reported to the National Research Council in 1949. "It can now step over its earlier limitations and use not only fresh unaltered minerals but also altered, oxidized minerals. Since most radioactive minerals contain three radioactive elements—U^{238}, U^{235}, and thorium—and all three produce leads of differing atomic weights and at differing rates, it is now possible by proper quantitative analysis to obtain three independent determinations of age in one and the same mineral. . . . As the most ancient minerals were formed as much as 2000 million years ago, such a threefold check on determinations of age is highly welcome in view of the enormous extrapolation backward in time.

"Summarizing the age evidence based on the radioactive minerals," Dr. Knopf continued, "we find that the oldest minerals are 2000 million years old. At least three fourths of geologic time, 1500 million years, had passed before the beginning of Cambrian time, when life other than algae first became abundant on this planet, or at least left records in the rocks. . . . Independent determinations agree . . . that the end of Early Permian time was 230 million years ago. The indicated length of the Paleozoic era is 300 million years, of the Mesozoic era 130 million years, and of Cenozoic time 70 million years."

One more ingredient was necessary for Osborn's success formula. The man who seemed most capable of fashioning it was induced to leave Chicago's Field Museum in 1909. Carl E. Akeley was an heir to the passionate interest in natural history and fossils that Amos Eaton had nurtured in upstate New York. One of the pupils and admirers of Eaton and James Hall was Henry A. Ward, a Rochester native possessed of a massive curiosity. Ward's collection of Genesee valley rocks, Lockport trilobites, and Chemung mammoth bones created such a clutter in his home and yard that, allegedly, his wife grew difficult. In order to relieve the chaos and restore domestic tranquility, Ward created his Natural Science Establishment, an institution that supplied schools, museums, and private collectors with anything from a pound of meteorite to a live electric eel or a mammoth skull. When P. T. Barnum's prize elephant, Jumbo, was mortally bashed by a locomotive, Barnum sent the carcass to Henry Ward to be stuffed and mounted.

During the 1880's, Ward's institution became the logical place to learn taxidermy, osteology, rock polishing, and fossil collecting—and it offered students the opportunity to get paid while they learned. It was the alma mater of W. T. Hornaday, director of the New York Zoological Society; of F. A. Lucas, director of the Brooklyn Museum and subsequently of the American Museum of Natural History; of A. B. Baker; and of Charles D. Walcott, secretary of the Smithsonian.

Carl Akeley was born on a farm northwest of Rochester in 1864. As an errand boy, and later as taxidermist, in Henry Ward's catacombs, he proved such an apt pupil that he joined the staff of the Milwaukee Museum in 1887. In 1895, he was invited down to Chicago, to the massive natural-history center

that the department-store millionaire, Marshall Field, was
building with and around the remnants of the Columbian Exposition. During the next nine years, Akeley began to fulfill
Charles Willson Peale's dream of dioramas that showed accurate reproductions of regional flora and fauna. He revived,
and mechanically improved, the Peale technique of preparing a
finely modeled statue of each animal and then stretching the
skin over it. He invented the cement gun, which became an
essential of taxidermy, and perfected methods for making lifelike plants, grasses, and trees out of fireproof materials.

The invitation to Akeley was one of Osborn's first acts as
president of the American Museum. In 1907, conferences with
"Uncle Pierpont" persuaded him to reject the offer to become
secretary of the Smithsonian Institution. His popularization
program, the steadily increasing crowds parading through the
exhibits on weekends, and his success with the Fifth Avenue,
Westchester, and Hudson River valley wealthy earned him election to presidency of the museum in 1908. He held this post for
twenty-five years and persuaded Rockefellers, Whitneys, War-
burgs, Dodges, as well as Morgans, to become trustees and, of
course, to lead the subscription lists for expeditions, purchases
of private collections, and construction of new buildings and
elaborate dioramas. "He ran the Museum on a personal basis,"
Cleveland E. Dodge told *The New Yorker*'s Geoffrey T. Hellman in 1968. "Until the Depression, J. P. Morgan, Ogden
Mills, my father and others would make up the annual deficit—
the wealthy trustees would give the annual dinners at their
houses. . . . Osborn was a very nice man. He had a very
gracious personality. He liked to have his own way. He got a lot
of these wealthy men on the board. They were a very loyal
bunch."

Some staff veterans disagreed with Mr. Dodge's impression of
Osborn's "very gracious personality." One woman, a curator,
called Osborn "exceedingly rude. He was impossible. . . . He
would knock you over getting into an elevator." Another described Osborn as "very lordly . . . [he had] a squirearch-
ical attitude. He had a belief in himself that you rarely en-

counter these days." However, none of the critics were paleontologists, and there is much evidence of interdepartmental fric-
tion and outspoken jealousy because of Osborn's continuing
devotion to his corps of dawnseekers.

The zeal for collecting fossils expanded to South America,
Canada, Mongolia, and Africa. On the eve of his ninetieth
birthday, Barnum Brown could boast, "The only major land
areas I haven't worked are Australia, New Zealand, Mada-
gascar and the South Sea Islands. The rest of the world has
been my hunting ground." He missed the Arctic and Antarctic
too, but only because he was too busy elsewhere. (His successor,
Edwin H. Colbert, was to fill part of this gap when he found, in
1969–70, reptilian fossils in Antarctica—which, incidentally,
gave key support to the concept of continental drift.)

In 1899 Brown was still sorting the bones from the 1898 dig
at Como Bluff when Osborn called him in and told him that
Princeton was preparing to send an expedition to southern
Argentina. There, since 1882, German, British, and Argentine
collectors had been digging dinosaur skeletons from the Cretaceous deserts of Patagonia. William Berryman Scott had
secured permission for a representative of the American Museum to accompany the Princetonians; Brown should be ready
to sail within the week.

John Bell Hatcher was codirector of the expedition. Hatcher
had lost his two-thousand-dollar-a-year post with Othniel Marsh
soon after the fundamentalist congressmen "got" Marsh. He
came east and found a teaching post at Princeton. Apparently,
the irregularity of pay checks from Marsh had forced Hatcher
to become a devoted participant in the all-night poker games at
frontier ranches and honky-tonks. Professor Hatcher started for
Patagonia "with just about enough pocket money to pay his bar
bill on the steamship voyage. But he promoted a 'few hands of
poker' with those Princeton dudes and landed at La Plata with
enough money to assure him of a two-year stay!" There is no
evidence that Barnum Brown was one of the contributors to
what is still snickeringly referred to as the Hatcher Patagonian
Fund.

By 1902, Brown was in eastern Montana. He explored the Judith and upper Missouri badlands for the next six summers. The specimens he sent to New York included two excellent skeletons of *Tyrannosaurus rex,* the largest of the carnivorous dinosaurs.

In 1909, Brown moved north to the Red Deer River valley of Alberta and extracted so many bones that the Canadian government called in the Sternbergs to get up there and "save some skeletons for Canada." India, Burma, the Aegean, all became Barnum Brown's hunting grounds. Roy Chapman Andrews, another Osborn star, called Brown "the lone wolf of explorers," adding, "I have known him to disappear from the Museum like the Vanishing American. No one knew where he had gone. . . . Inevitably his whereabouts was disclosed by a veritable avalanche of fossils descending on us in carload lots."

Twice during his extraordinary career as "the man who collected more dinosaurs than anybody on earth," Brown took leaves of absence to become an adviser on oil properties. "Seneca oil" and the other petroleum seepages that Amos Eaton, James Hall, and Charles Lyell had pondered about were, after 1850, transformed by an increasingly sophisticated technology into kerosene, lubricants, and finally gasoline and fuel oils. These fossil deposits provided the fuel for the reciprocal engine. The "gasoline revolution" radically changed the world's transportation systems, and drastically reduced transit times, but also it spewed concrete-and-steel grids across continents, encouraged the plow-ups that caused massive erosion and dust storms, fouled ocean estuaries and beaches, and spewed avalanches of poisonous gases into the atmosphere. Geologists and paleontologists became essential scouts for locating the subterranean pools where, millions of years ago, the fossil fuels had accumulated. During World War I, Brown was assigned to appraise oil properties for tax assessment by the Treasury Department. In the 1920's, he investigated the potentials for petroleum gushers in Abyssinia and, naturally, sent another "veritable avalanche" of African fossils back to American Museum.

In 1934, Brown pioneered another form of fossil prospecting when the Sinclair Refining Company—already using a dinosaur as its trademark symbol—financed his twenty-thousand-mile aerial survey across the Hayden-Powell-Marsh-Cope wonderlands. At the center of the monoplane's cabin, Weld Arnold of Harvard University operated one of the latest technological wonders, a twenty-four-inch Fairchild aerial survey camera. The nine hundred photos taken during the flights revealed "promising fossil areas in Triassic, Jurassic and Cretaceous strata previously thought to be completely explored. . . . a quarry of large dinosaur bones in a Cretaceous formation in which bones had not been previously reported . . . and unreported meteoric crater . . . and several new oil domes and structures not yet drilled."

Roy Chapman Andrews, a native of Beloit, Wisconsin, joined the American Museum staff in 1906 and, like Walter Granger, served his apprenticeship as a janitor and helper in the taxidermy department. By 1915 he was assistant curator of mammalogy and was obsessed with the idea that there might be fossil deposits in the Gobi Desert of Mongolia. After two exploratory trips to China's "Wild East," he was able to convince Osborn and the trustees to back the Central Asiatic Expedition. Andrews was given command of the expedition but, since he was skilled as a travel writer and public speaker but not as a paleontologist, Walter Granger was made codirector and was largely responsible for the immense collections of Mesozoic and Cenozoic fossils shipped to American Museum between 1922 and 1930. The expedition cost the museum's backers more than a million dollars, and thus was "the largest privately endowed land expedition ever to leave the U.S.A."; as such, it became a favorite subject for Sunday-supplement and magazine articles. The most publicized discovery was a clutch of twenty-five dinosaur eggs. Andrews used the eggs as exhibits during lectures and fund-raising campaigns in the winter of 1923–24; they netted $284,000 "for further Gobi excavations."

Still another Osborn star, George Gaylord Simpson, joined the staff in 1927 as assistant curator of vertebrate paleontology.

He soon proved he was as skillful in narration as he was in osteology. Collecting trips through the West and to Patagonia enabled him to trace the migratory and evolutionary patterns of prehistoric animals. His research added much detail to Othniel Marsh's studies of horse evolution. Simpson's book *Horses* is the best extant history and analysis of Equidae's evolution. Another facet of Simpson's genius was revealed in his delightful Patagonian journal, *Attending Marvels*. A third emerged in the scores of analytical, yet nonpedantic, papers and articles he published about the history of paleontological research in the United States. His definitive works include *Tempo and Mode in Evolution, The Major Features of Evolution,* and *Evolution and Geography*.

The museum's publication program kept pace with the popularization drive. Its monthly magazine, *Natural History*, was founded in 1897 and, although widely imitated, remained the best in the field. Scores of books written by staff members were published by special arrangement with major publishers. Photographs, bronze figurines, coloring books for children, "Museum Animal Theatres, sold only in sets of four for $1," and illustrated calendars became routine offerings.

When, on his seventy-fifth birthday, in 1933, Osborn testily relinquished the presidential suite, his paleontological collection was one of the world's largest. It continued to be as popular with visitors to Manhattan as the boat ride to the Statue of Liberty and the Goldman Band's summertime concerts on Central Park Mall.

Osborn did grumble now and then, but only to the veterans and always with a smile, about "those Carnegie fellows" having beaten them to the most stupendous dinosaur graveyard ever discovered. By 1900, the publicity given the great fossil hunts in the West had captured the interest of the Pittsburgh steel magnate Andrew Carnegie. He underwrote an expansion of the city's museum, and it was renamed the Carnegie Museum in his honor. Dr. W. J. Holland, the director, sent collectors out to Wyoming, Colorado, and Utah. One of them was Earl Douglass, a Minnesotan who had studied at Princeton and had been a favorite student of Dr. Scott.

During 1908 and 1909, after a bone hunt through the Uinta basin of Utah with Dr. Holland, Douglass searched for evidence of dinosaurs in the hogbacks of the Green River valley. Hayden and Powell had both noted these outcroppings during the 1870's and both had felt that there probably were big bones somewhere up in that Mesozoic turmoil.

On the afternoon of August 17, 1909, Douglass sighted "eight of the tailbones of a *Brontosaurus* in exact position" jutting from a saw-toothed ridge on the west side of the Green, about ten miles from the village of Vernal, Utah. Digging during the next few days convinced him that he had a major find. He went into Vernal, wrote Dr. Holland, bought camping equipment, and had most of the skeleton exposed when Holland arrived two weeks later.

The fact that the exposed dinosaur was one of the giants known as *Apatosaurus* soon became secondary. Evidence of other dinosaur remains extended all along the ridge. But they were in cameo and intaglio up and down the wall. Millions of years after the beasts had died, Holland deduced, the earth movements that created the contemporary contour of the Rockies had tilted the cliff up to an angle of about seventy degrees. The challenge, thus, was not to remove skeletons from a horizontal floor, but to dig a massive trench and then extract them from the semiperpendicular rock face.

Andrew Carnegie approved the Holland-Douglass plan for developing the Carnegie Dinosaur Quarry. Douglass directed the dig crews for fourteen years. More than three hundred tons of bones were extracted from the wall face and shipped to Pittsburgh.

A splendid skeleton that would create world-wide publicity for the Carnegie Museum and its benefactor was not from the Douglass dig but was a Wyoming discovery. In 1899, Bill Reed had again triumphed by digging out the complete skeleton of a vegetarian *Diplodocus* at Sheep Creek in Albany County,

Wyoming. It was a hitherto unknown species. John B. Hatcher, then curator of paleontology at the Carnegie Museum, shrewdly named it *Diplodocus carnegiei*. The steel titan was so flattered that he ordered plaster casts made from each of the monster's bones. When these were ready, Carnegie made a list of selected museums in Europe, plus the University Museum of Natural History in La Plata, Argentina, and the Museum of Mexico City, and he ordered Holland to have copies of "my dinosaur" made for each museum. Holland was to deliver each replica himself and personally supervise its installation. The task required most of a decade. Each formal presentation yielded medals for the Carnegie Museum or honorary degrees for Carnegie and Holland.

The dinosaur delivery from the Vernal locality was interrupted in 1914, when Douglass learned that the land on which the quarry was located was about to be thrown open by the federal government for homesteading. To protect the interests of science and its own investment, the Carnegie Museum filed a mineral claim to the quarry. But the claim was disallowed by the Department of the Interior on the grounds that dinosaur bones are not "minerals." The resulting hassle involved Charles D. Walcott, a disciple of Henry Ward who had recently succeeded to Joseph Henry's onetime post as secretary of the Smithsonian. Walcott, a paleontologist of vast accomplishment, realized the significance of the quarry to science and to the American people. It was he who, with Holland, took the matter to President Wilson; they pleaded so skillfully that the quarry became Dinosaur National Monument.

When the Carnegie Museum abandoned the dig in 1923, teams from the United States National Museum and the University of Utah began operations, working under temporary government permits. By the end of the following year, they too had extracted tons of skeletal parts and moved out, bringing to a close all excavating.

Meanwhile, in Pittsburgh, the three hundred tons of "dinobone" were being resurrected according to the meticulous formulas that Cope, Marsh, and Osborn had perfected between

1870 and 1910. Dr. Simpson, in 1933, wrote a very succinct description of the processes in the journal that would later be published as *Attending Marvels*. "Even when collecting is completed," he noted, "when the fossils are all securely packed in stout boxes, when local legal requirements have all been met, and when the shipment has actually reached the Museum—even then the work is only well begun. The rest takes place in the Museum, and invariably takes much longer than the time spent in the field. Preparators must clean each bone, removing the encasing matrix grain by grain. If soft, the bones must be carefully hardened, and if fragile, they must be reinforced with steel rods inside them or in some inconspicuous place along the outside. Fragments found scattered must be carefully matched, like a jig-saw puzzle, to see whether they will not fit together and make something more complete. Missing parts must be modeled in plaster, so far as this can be done with no possibility of error, by comparison with other similar specimens.

"When this work is done, the specimens must be studied. They are carefully compared with all others known and the name of the animal to which they belong is determined. If they are of some creature not known before, a new name must be given to them and a description published. New or otherwise particularly important things must be photographed or drawn. The age of the strata from which the fossils came must be determined. Detailed reports on the results of the whole expedition must be written and published.

"Finally, specimens must be selected and placed on exhibition. Iron supports have to be made to hold them, often a long and difficult task. The exhibition needs to be planned carefully, and comprehensible and enlightening labels composed and printed. Specimens not desired for exhibition must also be catalogued and then stored where they can be found readily when needed for study. In any collection there are numerous specimens, sometimes far the greater number, that are not exhibited. A specimen may have very great scientific value, and yet not have the properties of popular appeal, of completeness, of striking character, or of clarity in demonstrating some special point, that are requisite for appropriate exhibition in a large museum. The study collections of such an institution usually exceed the public exhibition both in bulk and in value."

A fastidious gourmet, Henry Fairfield Osborn in 1933 would have been horrified if anyone had dared to compare his museum to a gigantic smorgasbord. Yet, in effect, this is precisely what he had created along the two turreted blocks of Central Park West.

The intellectually nourishing and vibrantly colored feast was—and still is—the public display floors where absorbed visitors move in awe through the largest dinosaur collection in the world, the scores of anthropological and geological dioramas, and the 400,000 mammals, amphibians, and reptiles.

But there is much more nourishing material above and below these public areas. Fewer than ten per cent of the museum's 16,000,000 objects are on display at any one time. The other 15,400,000 are sorted in arched vaults and cabinets in the basements. Each piece is indexed and coded and is readily accessible for inspection by researchers and students.

The laboratories and workrooms of the osteologists, taxidermists, collectors, and anthropologists are on the upper floors. Here expeditions are planned, their treasures are analyzed, and the search for dawn facts reaches relentlessly toward sea bottoms, toward moon rocks and stratospheric dust.

Henry Fairfield Osborn died in 1935. His heritage was the achievement of that "dynamic museum to promote human understanding" that Charles Willson Peale had visualized in 1800.

19 | Pilgrimage

The world is too much with us; late and soon,
Getting and spending, we lay waste our powers:
Little we see in Nature that is ours.
—William Wordsworth

EACH book requires both quest and pilgrimage from its author. The quest for *The Dawnseekers* began to take shape at Laramie, Wyoming, in July, 1966. Again I walked across the University of Wyoming campus toward that concrete dinosaur guarding the museum entrance in order to impose on the patience of Dr. Paul McGrew, professor of paleontology. Paul is one of the world's greatest authorities on the evolution of the horse. I had a score of questions to ask about the jungle terrain of the Laramie Plain six hundred thousand centuries ago, when *Eohippus* scampered there. Also, I was bringing a handful of Equidae teeth that I had picked up on the Englewood and Sarasota beaches of Florida's Gulf Coast; had these been Florida natives, or had the teeth been washed in from elsewhere?

After answering my questions, Paul led me into the osteology laboratories. He stared for a moment at my handful of black and brown molars, opened a file drawer, fingered through the hundreds of horse teeth in it, then with casual deliberation identified my collection as "probably from *Merychippus* or one of the other early Nebraska grazers."

That afternoon, in a storage shed at his ranch home on the Laramie Plain, we examined slabs of rock about to be analyzed for possible evidence of bacterial life in the early Cambrian. To the north we could see the misty outline of the Cooper Lake badlands. There, a few years before, Dr. McGrew had found evidence that in 7000 B.C. horses still roamed in this land of their nativity. Obviously, thus, *Equus caballus* must have been known to the ancestors of the Cheyenne, Sioux, Crow, and Blackfoot. The McGrews' Morgan, quarter horse, and thoroughbred muzzled us for sugar cubes or carrot tidbits as we stared. Each of these three breeds had been created by man "less than a quarter second ago" on the time clock of their species' 600,000-century evolution.

By late afternoon another mutual interest had surfaced. Paul said, "My schedule is light tomorrow morning. Let's get out the camper, drive to Sherman Summit, and see how far we can follow the Union Pacific's original grade down the west slope."

On the eastern approach to the Dale Creek bridge in the spring of 1867, General Dan Casement's shovelmen had built a single-track embankment three hundred yards long and forty feet high across a meadow. Bottles of Nobel's deadly Patent Blasting Oil had been used to rip the right of way through a shale cliff to the bridge. Now there is barely enough room atop the embankment for a camper rig to cross into the rock cut. Dale Creek Bridge collapsed decades ago. The rock cut is too narrow for a turnaround. The only way out is to back those three hundred yards along the eroded top of the embankment.

But, first, there was still coffee in the Thermos, and chunks of turretine and cloudy blue agate glittered on the cliff wall. My Levi's pockets bulged with rock samples; there seemed to be room for two more. Paul lolled against the truck and watched me grunt up the rock face and pry out the nodules. He was staring at me so intently as I returned to the truck that I wondered how I had offended him. Then he grinned and drawled: "You're a confirmed nut about rocks. Nobody's yet written a history of paleontology in America." The grin spread. "You aren't a professional, so technically you're at the bottom of the barrel. You'd have to start from there. I think you should do it. I guarantee that it will be the most exciting book adventure you've ever tackled."

And it was! We were back in Paul's office by noon and taping his suggestions of authorities to interview. The research led from Boston to New Haven to Philadelphia to Washington, then west along Amos Eaton's grand canal route and on to the Hayden-Powell-Cope-Marsh glory lands.

There are, as all authors eventually discover, sputters of ego static during the efforts for communication. The acute compartmentalization and status struggles within and among contemporary professions have created a Tower of Babel effect in which the physicist, chemist, engineer, theologian, geologist, physician, and educator all use a special professional language. Many of them resent any effort to translate that special language into one accessible to the nonprofessional.

Both a county historian and a township historian in the mid-Genesee valley expressed complete ignorance about the identity of John Wesley Powell and insisted that he "could not have been born in Mount Morris because there wasn't a Methodist church there before the 1860's." Gracious Juliet Wolohan, associate librarian of the New York State Library, soon found a drawing of the Mount Morris Methodist Church in Barber and Howe's *Historical Collections of New York*. The book had been published in 1841.

There still isn't a historical marker at Powell's birthplace, next to the Methodist church site on that Genesee ridge. Nor is there one at the sites of the Michaelis and Peale digs in the Wallkill valley, or at the Eaton-Lyell-Jewell-Marsh search place in the Lockport cut. Amos Eaton, similarly forgotten in modern Chatham, is cited in textbooks merely as "a geologist who labored to make earth science accord with Genesis" and who "founded Rensselaer Institute."

Such indifference, however, was outweighed by the graciousness of scores of professionals. My file drawers overflowed with Xerox copies, tear sheets, correspondence, transcribed notes, and tape cassettes. The tops of reference books bristled with color-keyed "quote" and "restudy" flags. Thus it was soon time to get out into the field itself and begin my pilgrimage. No book is worth its pulp unless the writer has mastered his research. As Dr. Vernon Loggins of Columbia University always stressed and restressed: *Causal awareness is the ability to know what you are looking at.* Would I now know what I was looking at in the factory-strewn, macadam-crosshatched, garbage-fouled regions that, less than a century before, had been the sites of the Peale-Eaton-Lyell-Hayden-Powell-Cope-Marsh miracle digs?

Paul McGrew invited me to "bring your Levi's and hikers to Green River" and join his summer dig, "not far from the focal point of the Cope-Marsh fight." But first I wanted to re-experience the Sweetwater valley and South Pass. Jim Carpenter and Harold DelMonte had guided me up there several times, from the Burlington Grade and Mormon Handcarters' graves to Sioux Pass. How would Oregon Buttes, the red cliffs, and that amazing "dead end" of the Wind River Mountains' Precambrian appear, now that I had a little clearer notion of what the Old West really was and how its oldest old-timers have been resurrected?

Harold DelMonte owned the Hotel Noble at Lander, Wyoming. He was such a devoted history buff that each room had been given a theme, through paintings, hand-crafted furniture, and samples of Indian weaving. Harold's geological awareness showed in the display cases in the Noble's lobby, which held a dream lode of fossils, cloudy blue Sweetwater agates, prime Antelope Hills jade, and petrified woods from the cliffs of Bridger National Forest.

Jim Carpenter could not abide Lander's "city gewgaws." He had lived sixty-five of his seventy years at the edge of South Pass's sky-high prairie, and for thirty-five of them had served as deputy sheriff of Atlantic City and South Pass City. He knew more Indian, soldier, miner, and wagon-trail folklore about western Wyoming than any six experts. He still drove a four-wheel-drive truck so skillfully that, once or twice during each wander, we clocked antelope herds sprinting through the sagebrush. (At fifty miles per hour, they still disdainfully wagged white tails at us.)

The macadam ribbon of Wyoming 28 is the only highway that winds up Red Rocks Canyon to the South Pass causeway, where the mountain men, Oregon-or-bust pioneers, forty-niners, and Mormons all crossed the continental divide. Early in the 1960's the road had been regraded and widened when U.S.

Steel built a taconite plant on the "mountain of iron" that Ferdinand Hayden had identified and surveyed during the summer of 1871.

As we purred over the top of the scarlet cliffs, I hung so far out of the window that Harold nudged my back and chuckled, "I had a beagle once that acted just like that. He fell out and broke his tail." My excitement welled from the realization that, out of that green and tan and gray vista ahead, I could distinguish the thrust fault where a mighty river and millions of years of erosion had carved the gentle plain of South Pass between and across those cliffs of Precambrian, Paleozoic, Mesozoic, and Tertiary. As we turned into the forested dirt road that winds over to Atlantic City and Jim Carpenter's cabin, I thought I could distinguish the glimmer of fossil fish scales in the rocks. There just had to be dinosaur and mosasaur bones up there! (When I reported this discovery to Paul Mc-Grew two evenings later, there was an awkward silence and a sigh before Paul replied, "But why not?")

Jim's truck bounced us out past the collapsing cabins and trailing heaps of abandoned gold digs to the spires of Oregon Buttes. These form the south edge of South Pass and the north wall of the Red Desert. From their shadows we stared down across an oily shimmer of red, blue, green, and obsidian-black fire scars and lava domes.

Hell on Wheels pioneered the high-iron path across this Red Desert during 1868; the surveyors of U.S. 30, the Lincoln Highway, paralleled it a half century later. Almost due south of us twinkled the city of Rock Springs. Its sprawl of houses, rococo mansions, and mine heads began when Hayden and other geologists said, "Here is coal." Far to the southeast the tank farms and dinosaur statues at Sinclair symbolized the petroleum gushers that had been discovered through stratigraphic know-how all along the Medicine Bow range and throughout the Rattlesnake Hills. The gleam down toward Rawlins was the new desert community of Jeffrey City, where uranium was being processed. Vanadium mines roared nearby. All were products of paleontological skills.

An airplane broke the silence. I watched the macabre plume it spewed across the troposphere and, for the first time, realized, "That thing is up there because of fossil fuel. Autos and diesels run on the same explosive force. Nature produced petroleum drop by drop through eras of catastrophe. We are gobbling her whole supply in order to frenzy through a century—one-hundredth of a geologic second!—at insane, and basically useless, speeds."

Jim Carpenter was philosophizing, too. "Twenty years ago," he said, pointing north toward the Wind River cliffs, "I could walk four miles from here over to those bluffs and never touch dirt. Old Lady Nature paved this whole area with Sweetwater agates and petrified trees. Get some morning dew on 'em and they'd gleam like diamonds. So-o-o, along come the rock hounds, first with saddlebags, then with trucks, and finally with bulldozers. They yanked 'em all. Now you pay five dollars for a sniffy little piece of fossil twig in a rock shop, and a polished slab of agate no thicker than a Necco wafer costs eight or ten bucks. Those things are God's beautiful handiwork. He doesn't whack them out on a Detroit assembly line. Why aren't they protected by law?"

"Probably," Harold said, "because most people just don't give a tunket about natural beauty. Who was the poet who wrote, 'Little we see in Nature that is ours'? Wordsworth, I guess. Maybe they don't care because they don't understand. That's why they keep on fouling up the landscape with stuff like that litter down there."

He nodded down at the cans, bottles, cartons, and rope that a generation of picnickers, hunters, and agate pilferers had tossed into a gully. "The paper and rope and tin cans will take a century to self-destruct. But those bottles and aluminum beer cans will be around to intrigue archaeologists in 2970 A.D.— that is, if our little blue star is still twinkling then!"

The continental divide is a small hillock on Wyoming 28's roller-coaster toward Pacific Springs. The first trickle of the Little Sandy flows west from Pacific Springs' green hollow. Just beyond it, the Oregon Trail veered northwest toward Fort Hall and the chasms of the Snake River. But the forty-niners and the Mormons plodded southwest down the Sandy to its junction with the Green. In 1847, Brigham Young ordered a scow ferry built to cross the Green a few hundred yards from the Bridger Green cliffs, where the Sandy silts in. The forty-niners, the Pony Express, the telegraph construction crews, the Wells Fargo coaches, all used it. Hayden, King, Cope, Powell, Marsh, Hatcher, Sternberg, Osborn, and Barnum Brown came this way in portentous succession. Paul McGrew's summer students were camping on the west shore of the Green, a few hundred yards upstream.

Paul had ordered the river shore policed for a mile on each side of the camp; the tin-can, bottle, cigarette-box, and tire-tube refuse had been barreled and loaded in a truck for the next day's supply run to Green River City. "If a student can't learn to have respect for Nature," Paul explained, "then he or she isn't fit to be a paleontologist. The compounds that go into glass and tin cans and aluminum and synthetic rubber are mass-produced from Earth deposits that were discovered by applied paleontology. Professionally, then, paleontologists have a responsibility to help Nature balance the ecologic fouling by contemporary civilization. Anyone in this camp who starts littering automatically flunks the course."

"That's the dinner team," he explained a few moments later when six of the students hiked upstream, carrying fly rods. A half hour later they were back in camp, carrying two yard-long strings of rainbow trout. No chemist can compound a perfume so pleasing as batter-dipped rainbow trout sputtering in the "drippin's" of country-cure bacon.

We worked the black arroyos of Whiskey Basin during the next four days. This was Cope, Marsh, and Hatcher country, northeast of Kemmerer. We scrambled up and down cliffs, searching for big and little bones. We dug "sections" of pebbles and sand, poured them into paper sacks for laboratory study, and marked identification symbols on each bag. Each bone discovery brought a shout, scurry to the spot, a seminar. The ground area around the discovery was examined for fragments, possible skin impressions in the rock, and other circumstantial evidence. Then the piece was cautiously chipped out, examined in the rough, and carried to the truck. Back at camp, the swim in the Green that ended the workday didn't begin until all of the day's discoveries had been spread out on the portable tables, cleaned, and subjected to analysis and discussion.

Why, I brooded, wasn't the same awareness required for rockhounds? How much knowledge of South Pass's evolution had been lost when the looters moved in with bulldozers and vandalized that agatized forest? Why is the cultivation of geological awareness restricted to university majors and gradu-ate students? Dinosaurs and their jungle world fascinate most young children. Given some water colors and a chunk of wall, they will plot, diagram, and daub for days to produce a giddy, rainbowed mural of "The Reluctant Dragon" or "Dinosaur in Aunt Jane's Asparagus Patch." Why can't paleontology and Earth history be taught in elementary schools during those years of maximum receptivity? Where are the filmstrips and listening centers that would excite Jack, Jill, and Samantha about collecting fossils and would show the meticulous progres-sion from the tawny gleam in the rock to mounted display?

Mankind seems finally to be gaining an awareness of the ecological crises that have been created by technological greed since 1860. Fossils are the most vivid record of what happened when Nature caused species to change radically or to die out. The dinosaurs dominated our tiny blue planet for one hundred million years. Nature shifted gears; the dinosaurs vanished. Man has been around for a mere three million years. Technol-ogy has permitted us to become obsessed by giganticness—gigantic cities, gigantic bombs, gigantic dumps, gigantic smogs, gigantic ignorance of natural law. And still the thin rind of troposphere above us and the vastness of rock, water, and turmoil beneath us remain almost as mysterious as they were when Aristotle lectured on the hillsides of Athens.

The final shrine of my pilgrimage stands on Mesozoic cliffs a hundred miles south of the McGrew campsite. Habit and the population drifts across the West have favored east-west rather than north-south throughways. It is simpler to travel to the lower Green via Salt Lake City. Anyway, Sam and Lila Weller wanted to go out with me. Sam is an anomaly; his Zion City Bookstore features books rather than junk jewelry, paper clips, greeting cards, and souvenir towels. And his passion for Western Americana is as obvious as his thatch of red hair.

The highway east from Tabernacle Square soars up a tawny crevice in the Wasatch majesty, then dips into pastured valleys. We took the turnoff for Roosevelt and Vernal that for two hundred miles flirts and pirouettes with the Uintas. This mountain range still frustrates geologists because it "broke the pattern" and towers east to west instead of north to south. Nature's convulsions during the process helped the Green River gouge through the Uintas' eastern slopes, and belched up the ridge where Earl Douglass discovered the Carnegie Dinosaur Quarry.

At Vernal, the state of Utah has created Natural History State Museum. Dr. G. E. Untermann, the director, borrowed the casts of *Diplodocus carnegiei* from the Pittsburgh Museum and erected a weatherproof plastic reproduction of the huge skeleton in the field-house yard. He had already briefed me about the efforts of the Society of Vertebrate Paleontology at Yale to "get laws which will discourage the wanton destruction of fossil material in general," and frankly admitted that "enforcement is a joke, for who is going to see many of the people who prowl about in the wilderness? However, a bit of a crimp has been put in the California rock shops that were hauling this stuff off in semitrailer loads."

One section of the field house is set aside for amateur buffs like me. Bins of fossil fragments and such regional treasures as turretine and snowflake obsidian can be purchased by the pound. "The fossil material has been broken up too much by Nature to have scientific value," Dr. Untermann told me. Agatized chunks of dinosaur bone in the field-house bins cost fifty cents to two dollars a pound, depending on colors and quality. One pound of it had produced twenty-six of the blue-green brooch and ring insets that rock shops were retailing at eight to ten dollars per wafer.

Ten miles east of Vernal a macadam strip veers due north to dead end at a mud-trampled tan ridge that, like the jagged backbone of a dragon, juts up from the Green River shore. We stared 1,400,000 centuries up the road and didn't need the modest sign identifying this as Dinosaur National Monument. My research had transformed its bleakness into a cathedral more evocative of the Infinite God than any man-made structure.

During the Great Depression of the 1930's, the Works Progress Administration sent in hundreds of men to remove debris from the Carnegie Quarry, build a drainage system, and erect caretaker housing. The work was suspended during World War II. In 1953, Horace M. Albright, the visionary who energized the National Park Service, won appropriations for his plan to develop the north wall of the quarry in relief and "expose the remains as an in-place exhibit" under the supervision of Superintendent Jess Lombard and Theodore White.

Gilbert Stucker was one of the team of paleontological experts assigned to determine what the prospects were for a "shrine to the Mesozoic." "Some bone nubbins and the Carnegie Quarry charts were all we had to go on," Gil recalled. "The nubbins showed in a sandstone outcrop, one hundred eighty by thirty feet, which dipped at seventy-six degrees and extended twelve to fifteen feet back into the hill. . . . The proof lay within the rock itself. . . . . Air-powered jackhammers and rotary rock drills were used. Some fourteen hundred tons of rock were removed in the first few months. As bones showed, they were roughed out with small pneumatic chipping hammers, preparatory to the detailing with handtools."

A modernistic concrete-and-glass structure now arches over the quarry face. It will be extended until the ancient sandbar cemetery of bones pinches out. There aren't any juke joints or onion-reeking burger stands or tawdry souvenir shops on the premises. Diagrams and photographic murals in the anteroom explain the giant cameos that Mr. Stucker and his associates began in 1954.

The ridge itself forms the north side of the structure. Across its sandstone, the brown-and-black skeletons gleam in sun-dappled cameo. The fretwork of a *Diplodocus* tail cuts across a wall corner as though pointing at the enormous neck and skull of a *Camarasaurus*. Foot bones, vertebral columns, and drum-size centra rear in hypnotic patterns, not unlike a Jackson Pollock painting.

I grew up in parsonages and spent many childhood hours trying to pay attention to long sermons. As I stared at the dinosaur wall, I realized that my search for the dawnseekers had restored and expanded a faith in God that monotonous sermons and the behavior of the smugly pious had all but destroyed. Our little blue planet has had a destiny. It has played some sort of role in the unknown pattern of eternity. Here was Nature's Book of Revelations, a warning to audacious modern man, a reassurance for those who believe what every star-studded night so clearly implies: God is infinite and far too complex to create any species of animal in His Image on any tiny blue planet.

I recalled the quotation of the brass plate my wife had discovered in the floor of Westminster Abbey a few months before. It is Sir Charles Lyell's epitaph. But, we decided that afternoon, it is also a fitting memorial for every dawnseeker who pioneered awareness of the Earth. It reads: "Throughout a long and laborious life he sought the means of deciphering the fragmentary records of the earth's history in the patient investigation of the present order of Nature, enlarging the boundaries of knowledge and leaving on scientific thought an enduring influence. 'Lord, how great are Thy Works. Thy Thoughts are very deep.' "

The Age of the Earth:

GEOLOGIC TIME SCALE

The dawn of Earth is now estimated to have occurred approximately 4,600,000,000 years ago. Bishop Ussher placed the date of Creation at 4004 B.C.

Era	Period	Duration	Life Forms
Precambrian	Early Precambrian Late Precambrian	4,030,000,000 years	No fossils yet found Algae, sponges
Paleozoic	Cambrian	70,000,000	Trilobites, snails
	Ordovician	70,000,000	Clams, corals, crinoids, first fish
	Silurian	35,000,000	Scorpions, land plants
	Devonian	50,000,000	Sharks, insects, tree ferns
	Carboniferous	65,000,000	Amphibians, large insects, seed plants
	Permian	55,000,000	Reptiles, flies, conifers
Mesozoic	Triassic	35,000,000	First dinosaurs, ammonites, cycads
	Jurassic	54,000,000	Giant dinosaurs, archaic mammals and birds, modern insects such as bees
	Cretaceous	71,000,000	Marsupial and placental mammals, deciduous trees, grasses; dinosaurs and toothed birds die out

Era	Period		Duration	Life Forms
Cenozoic	Tertiary	Paleocene	11,000,000	Small mammals, earliest primates, modern toothless birds
		Eocene	16,000,000	
		Oligocene	12,000,000	Carnivores, rodents, mastodons, erect apes
		Miocene	19,000,000	Modern dogs and cats, mastodons in North America, true anthropoids in Old World
		Pliocene	4,500,000	First one-toed horse, transitional prehuman primates
	Quarternary	Pleistocene	2,500,000	Elephants, saber-toothed cats, camels, and horses in North America; rise of hominids
		Holocene	approximately last 10,000 years	Animals domesticated, plants cultivated, spread of man in Western Hemisphere, first cities slowly evolve

Acknowledgments

The co-operation of the following individuals and institutions was critical during the years of research preparatory to writing this book. I am, thus, deeply indebted to: Dr. Richard A. Bartlett, Professor of History, Florida State University, Tallahassee; Dr. Whitfield J. Bell, Jr., Librarian, American Philosophical Society, Philadelphia; Dr. W. E. Bigglestone, Archivist, Oberlin College, Oberlin, Ohio; Dr. Joseph L. Blau, then Acting Dean, Union Theological Seminary, New York City; Ruth E. Brown, Librarian, Academy of Natural Sciences, Philadelphia; F. Stuart Crawford, Editor, G. & C. Merriam Company, Springfield, Massachusetts; the librarians of the Cosmos Club, Washington, D.C.; Richard M. Dillon, Librarian, Sutro Library, San Francisco; Dr. Gene M. Gressley, Western History Research Center, University of Wyoming, Laramie; Harold Hacker, Librarian, Rochester (New York) Public Libraries; Clayton Hoagland, author and editor, Rutherford, New Jersey; Dr. Eleanore E. Larson, Professor of Education, University of Rochester, Rochester, New York; Cecil D. Lewis, Jr., Superintendent, Dinosaur National Monument, Vernal, Utah; Dr. Vernon A. Loescher, Senior Pastor, First Congregational Church, Chappaqua, New York; Clarence A. Lewis, then County Historian, Lockport, New York; Dr. Paul O. McGrew, Professor of Paleontology, University of Wyoming, Laramie; Blake McKelvey, then City Historian, Rochester, New York; Dr. Raymond W. Miller, historian and author, Washington, D.C.; Dr. A. R. Mortenson and his associate historians in the National

Park Service, Washington, D.C.; the librarians and stacks personnel at the excellent Americana collections of the University of Rochester, Rochester, New York; Dr. Effey M. Riley, Americana scholar, Rochester, New York; Gilbert F. Stucker, Vertebrate Paleontologist, The American Museum of Natural History, New York City; Alfred Stefferud, then Editor of *Friends Journal,* Philadelphia; Juliet Wolohan, then Curator of Colonial Americana, New York State Library, Albany.

Completion of the manuscript after a hiatus of invalidism was enabled by the diagnostic and surgical skills of Drs. George R. Lovell and Wheelock A. Southgate of the Genesee Hospital staff in Rochester, New York. Throughout the many months of the manuscript's preparation, Malcolm Reiss of Paul R. Reynolds, Inc., served patiently as agent and intermediary.

Nicholas Econopouly, George Gouskos, George Karpoutzis, and "Stelios" provided spiritual balm and practical assistance while the final drafts of the manuscript were being written in Greece. Jeannine Korman of the University of Rochester's College of Education prepared the final typescripts.

The patience and faith both of William B. Goodman, my editor at Harcourt Brace Jovanovich, and of my wife, Elizabeth Zimmermann Howard, have been extraordinary. During the years of my illness, Mr. Goodman sustained me with his sanative assurances. My wife, despite arduous duties as a principal at American Community Schools in Athens, proffered an abundance of love and objectivity that enabled me to finish the manuscript during 1974.

Bibliography

Paleontology's contributions to contemporary technology, folkways, and environment point up the need for a history of the profession that will be by and for scientists. *The Dawnseekers* was *not* meant to be a definitive history, but a factual, general-interest report on a neglected lode of important Americana. The author ardently hopes that the task of preparing an annotated history of American paleontology will soon be undertaken by the profession's scholars.

Similarly, the following bibliography is intended for those adults and young people who may wish to "dig" for more details about the pioneers of paleontological awareness and their impact on the technology and folkways of the United States of America.

The publications that proved most helpful during the years of research for *The Dawnseekers* were:

Books

Bartlett, Richard A. *Great Surveys of the American West*. Norman: University of Oklahoma Press, 1962.

Berry, William B. *Growth of a Prehistoric Time Scale*. San Francisco: W. H. Freeman, 1968.

Blau, Joseph L., ed. *Cornerstones of Religious Freedom in America*. New York: Harper & Row, 1964.

Carter, Paul A. *The Spiritual Crisis of the Gilded Age*. De Kalb: Northern Illinois University Press, 1971.

Clarke, James M. *James Hall of Albany, Geologist and Paleontologist*. Albany: Privately printed, 1921.

Colbert, Edwin H. *The Dinosaur Book*. New York: McGraw-Hill for The American Museum of Natural History, 1951.

————. *Men and Dinosaurs*. New York: Dutton, 1968.

Cook, James H. *Fifty Years on the Old Frontier as a Cowboy, Hunter, Guide, Scout and Ranchman*. 2nd ed. Norman: University of Oklahoma Press, 1957.

Coon, Carleton S. *The Origin of Races*. New York: Knopf, 1962.

Dillon, Richard H. *Meriwether Lewis; a Biography*. New York: Coward-McCann, 1965.

Eaton, Amos. *A Geologic and Agricultural Survey of the District Adjoining the Erie Canal in the State of New York.* Albany: Packard and Van Benthuysen, 1824.

Erskine, John. *The Moral Obligation to Be Intelligent.* Indianapolis: Bobbs-Merrill, 1915.

Fairchild, Herman L. *Geology of Western New York.* Rochester: Privately printed, 1925.

Gillispie, Charles C. *Genesis and Geology.* Cambridge, Mass.: Harvard University Press, 1951.

Godman, John D. "The Gigantic Mastodon," in *American Natural History.* Philadelphia: Stoddart & Atherton, 1828.

Haber, Francis C. *Age of the World; Moses to Darwin.* Baltimore: Johns Hopkins Press, 1959.

Hall, James. *Geology of New York: Part IV.* Albany: Carroll and Cook, 1842–43.

Homsher, Lola M., ed. *The South Pass, 1868: James Chisholm's Journal of the Wyoming Gold Rush.* Lincoln: University of Nebraska Press, 1960.

Howard, Robert West. *The Great Iron Trail: The Story of the First Transcontinental Railroad.* New York: Putnam's, 1962.

———. *The Horse in America.* Chicago: Follett, 1965.

Irvine, William. *Apes, Angels, and Victorians.* New York: McGraw-Hill, 1955.

Jepsen, Glenn L.; Simpson, George C.; and Mayr, Ernst, eds. *Genetics, Paleontology and Evolution.* Princeton, N.J.: Princeton University Press, 1949.

Lyell, Charles. *Travels in North America in the Years 1841–42.* New York: Wiley and Putnam, 1845.

McAllister, Ethel M. *Amos Eaton, Scientist and Educator, 1776–1842.* Philadelphia: University of Pennsylvania Press, 1941.

Marsh, Othniel C. *The Dinosaurs of North America; an Extract from the 16th Annual Report of the U.S. Geologic Survey, 1894–95.* Washington, D.C.: U.S. Government Printing Office, 1896.

Miller, Hugh. *The Testimony of the Rocks.* Boston: Gould & Lincoln, 1857.

Moorehead, Alan. *Darwin and the Beagle.* New York: Harper & Row, 1969.

Ogburn, Charlton. *The Forging of Our Continent.* New York: American Heritage, 1968.

Osborn, Henry Fairfield. *The Age of Mammals in Europe, Asia and North America.* New York: Macmillan, 1910.

———. *Cope: Master Naturalist.* Princeton, N.J.: Princeton University Press, 1931.

————. *Creative Education in School, College, University and Museum.* New York: Scribner's, 1927.

Peale, Rembrandt. *An Historical Disquisition on the Mammoth, or Great American Incognitum.* London: E. Lawrence, 1803.

Plate, Robert. *The Dinosaur Hunters.* New York: McKay, 1964.

Reingold, Nathan. *Science in Nineteenth-Century America; a Historical Disquisition.* New York: Hill & Wang, 1964.

Sellers, Charles C. *Charles Willson Peale.* New York: Scribner's, 1969.

Shor, Elizabeth N. *Fossils and Flies; the Life of a Compleat Scientist Samuel Wendell Williston.* Norman: University of Oklahoma Press, 1971.

Simpson, George Gaylord. *Attending Marvels; a Patagonian Journal.* New York: Macmillan, 1934.

————. *Horses.* New York: Oxford University Press, 1953.

————. *Life of the Past; an Introduction to Paleontology.* New Haven: Yale University Press, 1953.

Sternberg, Charles H. *The Life of a Fossil Hunter.* New York: Holt, 1909.

Ussher, James. *The Annals of the Old and New Testament.* London: E. Tyler for J. Crook, 1658.

White, Andrew D. *A History of the Warfare of Science with Theology in Christendom.* New York: Appleton, 1896.

Wright, Louis B. *Cultural Life of the American Colonies, 1607–1763.* New York: Harper, 1957.

Wyckoff, Jerome. *Rock, Time and Landforms.* New York: Harper & Row, 1966.

Zeuner, Friedrich E. *A History of Domesticated Animals.* New York: Harper & Row, 1964.

Periodicals, Monographs, and Unpublished Material

Barton, D. R. "Middlemen of the Dinosaur Resurrection." *Natural History,* May 1938, pp. 385–87.

————. "The Story of a Pioneer Bone Setter." *Natural History,* March 1938, pp. 224–27.

Bell, Whitfield J., Jr. "A Box of Old Bones: A Note on the Identification of the Mastodon, 1766–1806." *Proceedings of the American Philosophical Society* 93 (1949) :169–77.

Bishop, Francis M. "Letters to the Bloomington, Illinois, *Pantagraph* During Second Powell Expedition," with biography by Ralph B. Chamberlin. *Utah Historical Quarterly* 15 (1947) :241–44.

Bradley, George Y. "Journal During the First Powell Expedition on the Green and Colorado Rivers," with biography by William Culp Darrah. *Utah Historical Quarterly* 15 (1947) :31–54.

Brown, Barnum. "Flying for Dinosaurs." *Natural History*, September 1935, pp. 95–116.

———. "A Miocene Camel Bed-Ground." *Natural History*, November–December 1929, pp. 658–62.

———. "A Spine-Armored Saurian of the Past." *Natural History*, November–December 1932, pp. 493–96.

———. "Tyrannosaurus, a Cretaceous Carnivorous Dinosaur." *Scientific American*, 9 October 1915, pp. 322–23.

Dillon, Richard. "Stephen Long's Great American Desert." *Proceedings of the American Philosophical Society* 111 (1967) :93–108.

Grinnell, George B. "An Old Time Bone Hunt." *Natural History*, July–August 1923, pp. 329–36.

Hall, James, and Powell, John Wesley. Correspondence, 1881–94. New York State Library, Albany.

Hellman, Geoffrey T. "The American Museum, I." *The New Yorker*, 30 November 1968, pp. 68–150.

———. "The American Museum, II." *The New Yorker*, 7 December 1968, pp. 65–136.

Hoagland, Clayton. "They Gave Life to Bones." *Scientific Monthly*, February 1943, pp. 114–33.

Kohn, Lawrence A. "Charles Darwin's Chronic Ill Health." *Bulletin of the History of Medicine*, May–June 1963, pp. 239–56.

Leidy, Joseph. Correspondence, 1850–70. Academy of Natural Sciences, Philadelphia.

Lewis, Clarence O. "Ezekial Jewett Won Fame as Soldier and Geologist." Newspaper clipping dated 1 October 1959, but otherwise unidentified. Original is in the files of the Niagara County Historian, Lockport, New York.

———. "A Landmark of Niagara County." Unpublished essay about the birthplace of Othniel Marsh, in the files of the Niagara County Historian, Lockport, New York.

———. "The Morgan Affair." Monograph published by the Niagara County Historian, Lockport, New York (1966).

Matthew, W. D. "Early Days of Fossil Hunting in the High Plains." *Natural History*, September–October 1926, pp. 449–54.

McGrew, Paul O. "An Early Pleistocene (Blancan) Fauna from Nebraska." *Geological Series*, Field Museum of Natural History, 20 January 1944, pp. 33–66.

Osborn, Henry F. "Memorial of Samuel Wendell Williston." *Bulletin of the Geological Society of America*, 31 March 1919, pp. 66–76.

————, and Mook, Charles C. "Camarasaurus, Amphicoelias, and Other Sauropods of Cope." *Bulletin of the Geological Society of America*, 30 September 1919, pp. 379–88.

Proceedings of the Academy of Natural Sciences, Philadelphia, vols. 3–8 (1823–42).

Rezneck, Samuel. "Joseph Henry Learns Geology on the Erie Canal in 1826." *New York History*, January 1969, pp. 29–42.

Roseberry, C. R. "Before Cayuga." Adapted from a series of articles in the *Ithaca Journal* (1934), reprinted by the Ithaca Board of Education, Ithaca, New York (1950).

Simpson, George G. "The Beginnings of Vertebrate Paleontology in North America." *Proceedings of the American Philosophical Society* 86 (1942):130–88.

————. "The Discovery of Fossil Vertebrates in North America." *Journal of Paleontology*, January 1943, pp. 26–38.

————. "Hayden, Cope, and the Eocene of New Mexico." *Proceedings of the Academy of Natural Sciences* 103 (1951):1–21.

————, and Tobien, H. "The Rediscovery of Peale's Mastodon." *Proceedings of the American Philosophical Society* 98 (1954):279–81.

Stucker, Gilbert F. "Dinosaur Monument and the People." *Curator* 6 (1963):131–42.

————. "Harvester of the Past." *Nature*, November 1951, pp. 467–70.

————. "Hayden in the Badlands." *American West*, February 1967, pp. 40–85.

————. "Mountain of the Stone Fishes." *National Parks*, September 1966, pp. 4–9.

————. "Salvaging Fossils by Jet." *Curator* 4 (1961):332–40.

Index